# CHOLESTEROL & TRIGLYCERIDES

### Questions
### you
### have
### ... Answers
### you
### need

# Other Books From The People's Medical Society

Take This Book to the Hospital With You
How to Evaluate and Select a Nursing Home
Medicine on Trial
Medicare Made Easy
Your Medical Rights
Getting the Most for Your Medical Dollar
Take This Book to the Gynecologist With You
Blood Pressure: Questions You Have . . . Answers You Need
Your Heart: Questions You Have . . . Answers You Need
The Consumer's Guide to Medical Lingo
150 Ways to Be a Savvy Medical Consumer
Take This Book to the Pediatrician With You
100 Ways to Live to 100
Dial 800 for Health
Your Complete Medical Record
Arthritis: Questions You Have . . . Answers You Need
Diabetes: Questions You Have . . . Answers You Need
Prostate: Questions You Have . . . Answers You Need
Vitamins and Minerals: Questions You Have . . . Answers You Need
Good Operations—Bad Operations
The Complete Book of Relaxation Techniques
Test Yourself for Maximum Health
Misdiagnosis: Woman As a Disease
Yoga Made Easy
Massage Made Easy
Hearing Loss: Questions You Have . . . Answers You Need
Asthma: Questions You Have . . . Answers You Need
Depression: Questions You Have . . . Answers You Need
Back Pain: Questions You Have . . . Answers You Need
Stroke: Questions You Have . . . Answers You Need
Self-Care: Your Family Guide to Symptoms and How to Treat Them
So You're Going to Be a Mother

# CHOLESTEROL & TRIGLYCERIDES

## Questions
## you
## have
## ...Answers
## you
## need

By Ellen Moyer

≡People's Medical Society®

*Allentown, Pennsylvania*

**The People's Medical Society** is a nonprofit consumer health organization dedicated to the principles of better, more responsive and less expensive medical care. Organized in 1983, the People's Medical Society puts previously unavailable medical information into the hands of consumers so that they can make informed decisions about their own health care.

Membership in the People's Medical Society is $20 a year and includes a subscription to the *People's Medical Society Newsletter.* For information, write to the People's Medical Society, 462 Walnut Street, Allentown, PA 18102, or call 610-770-1670.

This and other People's Medical Society publications are available for quantity purchase at discount. Contact the People's Medical Society for details.

© 1995 by the People's Medical Society
Printed in the United States of America

Library of Congress Cataloging-in-Publication Data
Moyer, Ellen.
    Cholesterol & triglycerides : questions you have—
answers you need / by Ellen Moyer.
       p.   cm.
    Includes bibliographical references and index.
    ISBN 1-882606-51-5
    1. Hypercholesteremia—Popular works.  2. Hyperlipidemia—
Popular works.  3. Lipids in human nutrition—Popular works.
I. Title.
RC632.H83M69   1995
616.1'205—dc20                        95-24806
                                      CIP

2 3 4 5 6 7 8 9 0
First printing, September 1995

# CONTENTS

# INTRODUCTION

Oh no! Not another book on cholesterol. Haven't they written everything there is to know about the subject?

Frankly, I thought they had until we began to look at the latest medical research. What we found was significant. In the last few years, major findings have been made about the connection between heart disease and elevated, "bad" cholesterol. Even more has been learned about the so-called good cholesterol. And a whole body of knowledge is emerging about triglycerides and their impact on heart disease.

In addition, a major debate is going on between cholesterol and heart-disease researchers on the actual risk of developing heart disease as a result of high cholesterol levels. It seems that we in the general public have been hearing only the "down" side of the argument, while little is being said openly about the views of some experts who say high cholesterol levels pose less of a threat, in and of themselves, than we have been led to believe.

But there is still another reason for "another" book on the subject. From our mail and our conversations with consumers all across the country, it is quite clear that most people are totally confused about what and whom to believe on the subjects of cholesterol and triglycerides. Every day newspapers are filled with what seem to be contradictory articles, based on even more contradictory medical research, on cholesterol and triglycerides.

We felt it was time to set the record straight. Unlike most other books you may find on the subject, we are not trying to sell a viewpoint. Instead, we are passing along the facts, as they have been set forth in the most reliable medical studies. We discuss the known, the unknown and the gray areas of the medical world's knowledge on cholesterol and triglycerides. And we do it in a readable question-and-answer format—a format that the People's Medical Society has successfully employed since our founding in 1983.

What makes the subject of cholesterol and triglycerides interesting is that it has become an industry. There are probably more cookbooks, guides, television programs and printed articles about lowering cholesterol levels than on any other health subject. And yet high cholesterol is not a disease—or even a condition! Cholesterol and triglyceride counts are nothing more than measurements. There is *no* evidence that suggests a high count of either measurement is a guarantee of developing heart disease. There *is* evidence that, in connection with other risk factors, high cholesterol and triglyceride counts raise heart-disease risk. But does this evidence truly warrant the cholesterol industry that has developed? We have attempted to answer that question along with scores of others in the pages that follow.

So, yes, this is another book on cholesterol. However, we believe that it offers the reader the opportunity to weigh earlier findings with the latest research. And, in so doing, it helps make you a more informed, active participant in your own health care.

**Charles B. Inlander**
President, People's Medical Society

# CHOLESTEROL & TRIGLYCERIDES

**Questions
you
have
. . . Answers
you
need**

*Terms printed in boldface can be found in the glossary, beginning on page 169. Only the first mention of the word in the text will be boldfaced.*

*We have tried to use male and female pronouns in an egalitarian manner throughout the book. Any imbalance in usage has been in the interest of readability.*

# 1 CHOLESTEROL BASICS

**Q:** I know that having high **cholesterol** is supposed to be bad for me, and that cholesterol is somehow related to fat, but beyond that, I'm in the dark. So, what exactly is cholesterol?

**A:** Cholesterol is a white, waxy substance found naturally throughout the body, including the blood, and is essential for good health.

Technically, cholesterol is not a fat but a closely related substance. It belongs to a class of compounds called **sterols**. But it's often called a blood fat, or **lipid**.

Like any wax or fat, cholesterol does not dissolve in water. So in the blood, cholesterol is carried around in an envelope of protein. This cholesterol-protein package, called a **lipoprotein** (of which there are several kinds), does stay soluble in the watery serum portion of the blood.

That's why, when doctors measure cholesterol, they often do it in two ways: First, they measure total cholesterol—all the cholesterol in your blood. And second, they measure each of the different lipoproteins in your blood that, together, make up your total cholesterol. Each amount is measured as milligrams per deciliter (mg/dL). We'll explain more about these different lipoproteins in a bit.

**Q:** What about **triglycerides**? Aren't they related to cholesterol?

**A:** Not really. Triglycerides are also fats, or lipids, but they have a chemical structure that is different from cholesterol's. A triglyceride molecule contains three chains of fat—hence, the "tri" prefix. And instead of being attached to a protein molecule, it is attached to a glycerol (alcohol) molecule, which, like protein, is soluble in the watery serum part of the blood.

**Q:** Does a triglyceride molecule contain cholesterol?

**A:** No, it is a completely different fat.

**Q:** Where do cholesterol and triglycerides come from?

**A:** You can get both cholesterol and triglycerides from dietary sources—foods.
Cholesterol comes only from the animal foods you eat. Triglycerides are found in both animal fats and plant oils. Organ meats, such as liver and brains, contain lots of cholesterol, as do egg yolks, shrimp and lobster. Other meats, butter and whole milk also contain substantial amounts.

However, you may be surprised to learn that your liver actually makes about two-thirds of the cholesterol and some of the triglycerides in your body.

**Q:** **How does the liver make cholesterol and triglycerides?**

**A:** To make cholesterol, it uses the fats you eat, primarily **saturated fats**, such as butter or lard, which are hard at room temperature. That's one reason a high-saturated-fat diet tends to lead to high blood-cholesterol levels. In the case of triglycerides, both alcohol and sugar increase the liver's production.

**Q:** **What exactly does cholesterol do? You said earlier it's essential for good health.**

**A:** Cholesterol does serve a useful purpose. It is needed by the body to make hormones, including sex hormones and adrenal hormones, and to make vitamin D. It's also needed to produce bile acids, which aid fat absorption in the small intestine, and to build cells, especially the fatty membranes that enclose cells and the structures within cells.

**Q:** **And what do triglycerides do?**

**A:** They provide fats that are either burned for energy or deposited into the body's fat stores—the fat found under your skin and around your middle—to be used later for energy.

**Q:** It sounds like we need this stuff. Then why is having high cholesterol considered to be so bad?

**A:** High cholesterol, by itself, isn't a disease. It's simply a measurement that's used to assess your risk for disease. And certainly not everyone with high cholesterol develops health problems.

On the other hand, studies show that, in general, people with high blood levels of cholesterol are more likely than people with normal or low levels to have some of the cholesterol stick to the inner surfaces of blood vessels, causing **coronary-artery disease** and **atherosclerosis**. We explain in Chapter 4 why having very *low* cholesterol might also be a problem for some people.

**Q:** Before we go on—what's considered a high blood level of cholesterol?

**A:** These days, most doctors consider any level over 200 mg/dL to be a potential problem, depending on other risk factors you may also have. And levels over 240 mg/dL definitely put you at higher risk of developing atherosclerosis.

**Q:** And what is coronary-artery disease? Heart disease?

**A:** It's a specific kind of heart disease. Coronary-artery disease is atherosclerosis of the coronary arteries, the spaghetti-size arteries that deliver blood to the muscles of the heart.

**Q:** Atherosclerosis. That's hardening of the arteries, right?

**A:** Well, almost. Atherosclerosis is a medical term that refers specifically to the gradual buildup of fatty deposits, called **plaque**, on the inside walls of the arteries.

A similar word, **arteriosclerosis**, literally means "hardening of the arteries." It is a broad term used to cover a variety of diseases, including atherosclerosis, that lead to abnormal thickening and hardening of artery walls.

**Q:** But as for *athero*sclerosis—how does cholesterol end up on artery walls?

**A:** Only one type of cholesterol, called **low-density lipoprotein** (**LDL** for short), ends up on artery walls in any appreciable amounts. In studies, people with high levels of LDL have an increased risk for atherosclerosis. On the other hand, people with high amounts of another type of lipoprotein, **high-density lipoprotein** (**HDL**), have a reduced risk for atherosclerosis.

## LDL CHOLESTEROL

**Q:** **Is LDL cholesterol the kind that's called "bad" cholesterol?**

**A:** Yes. LDL contains lots of cholesterol. Its job in the body is to transport cholesterol from the liver to the cells. If there is more cholesterol available than cells can take up and use, however, LDL ends up circulating in the bloodstream until, eventually, it sticks to artery walls.

**Q:** **Does it stick just anywhere?**

**A:** It tends to stick in places where the innermost layer of cells, a protective barrier of **endothelial cells**, has been damaged. It sticks to the walls of arteries that have microscopic tears or rough spots from high blood pressure, toxins from cigarette smoke, or bacterial or viral infections, and in places where blood flow is turbulent because the artery branches or because circulation has been altered by bypass surgery.

**Q:** **So that's all? It just piles up on artery walls and eventually blocks them off?**

**A:** No, the process is fairly complicated, at least by microscopic standards, explains Peter Kwiterovich, M.D., chief of the lipid research unit at Johns Hopkins University School of Medicine, in Baltimore.

The LDL cholesterol, which is a relatively small molecule, squeezes through the spaces between endothelial cells until it is inside the wall of the artery. There, it reacts with oxygen in a chemical process called **oxidation** (which, incidentally, certain vitamins can help stop; we talk more about the protective role of antioxidant vitamins in Chapter 5).

Once the LDL cholesterol is oxidized, the process really takes off. The oxidized LDL tends to draw more LDL toward it, like a magnet. Then **macrophages**, immune-system cells designed to clean up messes, move in and gobble up the oxidized LDL cholesterol.

**Q:** So these cells clean up this mess?

**A:** No, in this particular situation they seem to just make it worse. "As long as LDL levels remain high, macrophages literally stuff themselves to death eating oxidized LDL," Kwiterovich says. In the process they turn into **foam cells**—large, fluffy cells that, under a microscope, really do look foamy. The foam cells die, spilling their contents within the artery wall, causing cracks in the artery wall that attract **platelets**—small, disk-shaped structures involved in blood coagulation. The platelets secrete chemicals that attract smooth muscle cells that proliferate and form scar tissue. Later, calcium in the bloodstream may cover the whole mess with a hard shell.

**Q:** So it sounds like what we're calling plaque—cholesterol deposits in arteries—contains a lot of different things, not just cholesterol.

**A:** You're right. The plaque consists of cholesterol, other lipids, fibrous materials and cellular debris, and minerals.

In fact, there are two different kinds of plaque. The one we've just described is called *active* plaque, and while it has the greatest potential for causing problems because it's most likely to cause blood clots, it's also the only type that seems able to actually shrink in size, given the proper conditions. We talk about the regression of heart disease in Chapter 5.

The other type of plaque is mostly cholesterol encased in scar tissue. It may be older than active plaque and has no signs of macrophage activity. It usually doesn't cause problems, but it doesn't shrink, either.

**Q:** What's considered a good level of LDL to have?

**A:** Below 130 mg/dL is considered desirable. Borderline high is 130 to 159 mg/dL, and high is 150 mg/dL or above. People who have heart disease are advised to try to keep their LDL cholesterol levels at 100 mg/dL or lower.

## TRIGLYCERIDES

**Q:** Do triglycerides build up on artery walls?

**A:** There is very little evidence that triglycerides build up, or deposit, on artery walls.


**Q:** But isn't a high blood level of triglycerides potentially harmful?

**A:** A high blood level of triglycerides, by itself, may not be so bad. But high triglyceride levels frequently go hand-in-hand with low levels of so-called good HDL cholesterol, which we discuss in a minute, and often, with high levels of LDL cholesterol.

One study, the Helsinki Heart Study, found that people with high blood triglyceride levels alone—no other risk factors—had about a 50 percent increased risk for coronary-artery disease, compared with people with normal levels.

However, they had a *threefold* risk for coronary-artery disease when they had both high blood triglyceride levels and low levels of HDL.

And their risk for developing coronary-artery disease was five times greater compared with people with desirable levels of triglycerides when they also had at least borderline high blood pressure (140/90 mmHg or above.) So triglycerides may do most of their dirty work when other risk factors are present.

Further, recent studies suggest that high triglyceride levels make blood more readily form clots, another important factor in clogged arteries.

**Q:** **What is considered a high level of triglycerides?**

**A:** Levels above 500 mg/dL are considered very high, and 250 to 500 is borderline high. For most people, levels below 200 are considered normal. We have more to say about ranges of triglycerides in Chapter 3.

## HDL CHOLESTEROL

**Q:** **So now I know about "bad" cholesterol. What about the good stuff, the kind that's supposed to help prevent plaque buildup?**

**A:** You're talking about high-density lipoprotein (HDL), which contains the least amount of cholesterol. HDL is considered good because it can remove cholesterol from cells and carry it back to the liver, where it can be excreted from the body via bile acids secreted into the intestines. So, high levels of HDL indicate that cholesterol is being removed from the body.

Components in HDL also help reduce blood clotting and blood-vessel constriction. That means people with high levels of HDL are less likely to form blood clots in their arteries or to have blood-vessel spasms that can squeeze off blood supply.

In studies, people with high levels of HDL show a lower risk of developing atherosclerosis. And people with low or even borderline levels of HDL may have increased risk, even when total cholesterol levels are normal. We'll talk more about cholesterol levels in Chapter 3.

Overall, studies find that for each 1 percent increase in HDL cholesterol, there is about a 3 percent decrease in your risk of coronary heart disease.

**Q:** **What's considered a good level of HDL to have?**

**A:** At least 30 mg/dL if your total cholesterol is very low—180 mg/dL or less. For most people, levels of 35 mg/dL or higher are best. Some people's HDL levels reach 75 mg/dL or higher. In the average man, HDL cholesterol levels range from 40 to 50 mg/dL; in the average woman, from 50 to 60.

## OTHER RISK FACTORS FOR HEART DISEASE

**Q:** **Aren't there other things besides high cholesterol that are just as bad for your heart? Like smoking, for instance?**

**A:** Yes. Keep in mind that high cholesterol is only one among many risk factors for heart disease and that it's really only a risk factor, not a disease in itself. For some people, unless cholesterol is very high, it is not even their strongest risk factor for heart disease.

**Q:** What are the other risk factors?

**A:** Doctors divide them into two categories: those that can't be changed and those you can do something about.

**Q:** What are the risk factors you can't change?

**A:** Your age, being male rather than female, and having a family history of premature heart disease—before age 55 in men and age 65 in women.

**Q:** How does your age affect your risk for heart disease?

**A:** The older you are, the greater your risk, no matter what your cholesterol level.

Although risk for heart disease increases continuously with age, heart attacks, angina and other signs of atherosclerosis are relatively rare until age 45 in men and age 55 in women. After that, they increase progressively. For instance, the chance that a 62-year-old man will die of heart disease in the next year is 500 times that for a 22-year-old man. High-cholesterol problems in older people are discussed in Chapter 4.

**Q:** **How does gender affect risk for heart disease?**

**A:** Men are at higher risk because, until menopause, a woman is protected from heart disease by the female hormone estrogen. That's why men in their 40s are four times more likely to die of heart disease than are women of the same age. After menopause, however, a woman's risk increases; but it's still never as high as a man's risk. At age 70, a man is still twice as likely as a woman to die from heart disease. We talk more about women and heart disease in Chapter 4.

**Q:** **And what if you have a family history of premature heart disease?**

**A:** Your risk for heart disease does increase, but just how much is uncertain. Some studies show as much as a twelvefold increase in risk, others only about 25 percent. A study by researchers at Harvard Medical School found that your risk for heart disease doubles if either your mother or father had a heart attack before age 70, and that your risk increases even more the younger your parent was when he or she had the heart attack. Inherited cholesterol problems are presented in more detail in Chapter 4.

**Q:** **What are the risk factors you can change?**

**A:** These include cigarette smoking, high blood pressure, lack of exercise, obesity, diabetes, stress and, of course, high cholesterol.

**Q:** **In that order? Are they ranked according to how much of a risk they present?**

**A:** Only generally. Smoking, high cholesterol, uncontrolled high blood pressure and lack of exercise are considered major risks, but all kinds of variables come into play—how long you've smoked and how much, for instance, or how high your cholesterol or blood pressure is. Same goes for the minor risk factors: obesity, stress and diabetes.

**Q:** **How do these other risk factors stack up against high cholesterol when it comes to a person's risk of developing heart disease?**

**A:** Here again, many variables must be considered. It's possible, however, to make some generalizations. For instance, smokers are 2½ times more likely than nonsmokers to have a heart attack. That risk is comparable to what your risk would be if your cholesterol level was 300 mg/dL or so. We talk more about smoking in Chapter 7.

**Q:** **And what about high blood pressure? How does it increase your risk for heart disease?**

**A:** As your blood pressure rises from normal (120/70 mmHg) to between 140/90 and 159/99, your risk of dying of a heart attack or stroke doubles. If it gets to the 160/100 to 179/109 mmHg range, your risk triples.

# Q: And diabetes?

A: Everything else being equal, the risk of heart disease is increased five times in a diabetic woman and two times in a man. The jury is still out, however, on the question of whether careful control of blood-sugar levels can decrease these cardio-vascular risks.

# Q: And obesity?

A: It's excess fat, not body weight alone, that seems to increase risk. Excess fat is best measured by something called Body Mass Index (BMI), which is, simply, your weight in kilograms divided by the square of your height in meters.

Here's a quick conversion to help you determine your BMI: First determine your weight in kilograms. One kilogram is equal to 2.2 pounds, so you must divide your weight in pounds by 2.2. Next you must determine your height in meters. One inch is equal to .0254 meters, so you must multiply your height in inches by .0254. Now take your height in meters and square it—that is, multiply it by itself. Finally, divide your weight in kilograms by the square of your height in meters. The resulting number is your BMI.

In one study, women with a BMI greater than or equal to 29 had almost double the risk of heart disease compared with women with a BMI of less than 21. We talk more about body weight and heart disease in Chapter 7.

**Q:** How does high cholesterol interact with these other risk factors?

**A:** The risks are multiplied by any combination of risk factors, including cholesterol. For instance, if you simply smoke cigarettes, you have a 4.6 chance in 100 of having a heart attack within the next eight years. If you smoke and have high cholesterol, your chances are 6.4 in 100. And if you also have high blood pressure, your chances are about 9.5 in 100.

That's why anyone with high cholesterol needs to look at his other risk factors for heart disease so he can develop a plan to reduce his overall risk, not just his high cholesterol.

It's true, there's nothing you can do about growing older, being male or having a genetic risk for heart disease. But the more unchangeable risk factors you have, the more likely you are to benefit, perhaps in the near future, by working on the risk factors you can change.

In Chapter 2, we look at studies that show just how much of a risk for heart disease high cholesterol has proved to be.

# 2 HIGH CHOLESTEROL AND HEART DISEASE

## STUDIES LINKING HIGH CHOLESTEROL AND ATHEROSCLEROSIS

**Q:** You said earlier that studies show an association between high cholesterol and heart disease. What can you tell me about these studies?

**A:** Both animal studies and studies of people suggest a strong link between high cholesterol and atherosclerosis, or heart disease.

In 1908, a Russian researcher first noted the connection in rabbits. Later, studies of monkeys demonstrated the direct relationship between cholesterol and saturated fat in the diet and cholesterol levels in the blood and the development of atherosclerosis.

In one such study, for instance, a group of monkeys was fed a typical American diet high in total fat, saturated fat and cholesterol, while another group was fed a prudent diet lower in fat. The animals on the typical American diet had significantly higher levels of blood cholesterol and much more atherosclerosis in their coronary arteries than those on the prudent diet.

# Q: What about human studies?

**A:** Human studies have pretty much confirmed the connection between high blood-cholesterol levels and increased risk for atherosclerosis. They've also confirmed the link between diets high in saturated fat and high cholesterol levels.

For instance, as early as the 1960s, medical researchers determined that not all countries had the same amount of atherosclerosis in their populations. Countries with the most heart disease, such as the United States and Norway, had higher average blood-cholesterol levels and consumed more fat than countries with a low rate of heart disease, such as Japan.

# Q: Any more recent studies?

**A:** A number of major studies have looked at the link between blood levels of cholesterol and risk of heart disease in humans.

One of these, the Multiple Risk Factor Intervention Trial (MRFIT), which ended in 1986, found a strong relationship between blood-cholesterol levels and death from coronary heart disease. For six years the study followed 361,662 American men, ages 35 to 57, with no history of heart attack.

This study found that the risk of death from coronary heart disease begins to increase gradually at blood-cholesterol levels of 180 mg/dL, accelerates at about 200 mg/dL, doubles at about 220 mg/dL (in comparison to 180 mg/dL and below) and triples at about 245 mg/dL. We talk more about cholesterol levels and ranges in Chapter 3.

In this study the bad effects of high cholesterol were isolated from other major risk factors, such as high blood pressure and cigarette smoking. Doing that helped establish that high cholesterol, by itself, is a risk factor for atherosclerosis, at least in middle-aged men.

**Q:** **Any other important studies?**

**A:** One worth mentioning is the world-famous Framingham Heart Study, which found that, at any age, heart-attack rates rise 2 percent for each 1 percent increase in blood cholesterol, starting at about 200 mg/dL. People with blood-cholesterol levels of 200 mg/dL or lower had about a 10 percent risk of coronary artery disease, while people with levels of 240 mg/dL or higher had an 18 percent risk.

That means 1 out of 10 people with cholesterol lower than 200 mg/dL, and about 2 out of 10 people with cholesterol higher than 240 mg/dL will develop heart disease.

**Q:** **Have any studies found no connection between high cholesterol levels and heart disease?**

**A:** There are some that show only a weak link. Just how strong a connection is found depends on lots of factors: the people being studied, the length of the study and the means used to determine heart disease. Links between cholesterol and heart disease are weaker in women, especially premenopausal women, and in both men and women older than age 70. Also, researchers now know that total cholesterol is not as good a predictor of heart disease as are some other cholesterol measurements, such as HDL-to-LDL ratio, which we talk about more in Chapter 3.

**Q:** Wasn't there a recent study that showed that people over a certain age don't have to worry if their cholesterol is high?

**A:** Yes. A study by Yale University researchers found that high cholesterol may not be as good a predictor of heart disease in people age 70 or older as it is in younger people. We talk more about what older people need to know about high cholesterol in Chapter 4.

## PROOF THAT LOWER CHOLESTEROL MEANS LESS HEART DISEASE

**Q:** Okay, you've presented evidence to show that high blood levels of cholesterol are associated with an increased risk for heart disease. But is there proof that lowering cholesterol levels reduces a person's risk for heart disease?

**A:** Yes. Most of the 20 or more studies that have looked at this question showed a significant reduction in the incidence of coronary heart disease after blood-cholesterol levels were lowered, either with diet alone or with diet and drugs.

Many of the studies using cholesterol-lowering drugs failed to show a reduction in *total* deaths, however, and some showed an increase. That's a disturbing finding we discuss in more detail below and in Chapter 4.

**Q:** **What can you tell me about the studies that show that lowering cholesterol reduces a person's risk for heart disease?**

**A:** One of the earliest, the Lipid Research Clinics Coronary Primary Prevention Trial, included 3,806 healthy middle-aged men with high levels of total blood cholesterol and LDL cholesterol; they were followed for seven years during the 1970s.

All the men followed a reduced-fat diet and also took either a cholesterol-lowering drug called **cholestyramine** or a placebo (inactive drug) that looked just like it. The men taking the cholestyramine decreased their total cholesterol 11.8 percent and their LDL cholesterol 18.9 percent. They also had a slight increase in HDL cholesterol. The men taking the placebo had reductions of 5 percent in total cholesterol and 8.6 percent in LDL.

Deaths from heart attacks were almost 20 percent less in the men taking cholestyramine compared with the men taking the placebo. At cholesterol levels above 200 mg/dL, for each 1 percent reduction of blood cholesterol and LDL cholesterol there was a 2 percent reduction in coronary heart disease. There was also a similar reduction in **angina pectoris** and coronary-artery bypass surgery. And **electrocardiography** tests, which measure and record electrical activity in the heart, detected less coronary artery blockages in the people whose cholesterol was lowered.

**Q:** Do other studies show similar results?

**A:** Yes, they tend to.
    Another notable study, with results published in 1987, was the Helsinki Heart Study. This five-year trial included more than 4,000 healthy, middle-aged Finnish men. The men did not have evidence of heart disease, but they did have high blood levels of LDL cholesterol and **very-low-density lipoprotein (VLDL)** cholesterol. They received either the cholesterol-lowering drug **gemfibrozil (Lopid)** or a placebo.

In the group taking the drug, there was significant reduction in total cholesterol (10 percent) and LDL cholesterol (11 percent) and an 11 percent increase in HDL. Compared with a control group taking a placebo, there were 34 percent fewer new cases of angina or heart attacks in the men taking the drug.

## CHOLESTEROL LOWERING IN PEOPLE WITH HEART DISEASE

**Q:** What about people who already have signs of atherosclerosis? Do studies show that lowering cholesterol helps them avoid a heart attack?

**A:** Most do. For instance, the Coronary Drug Project, done in the early and mid-1970s, used two medications: **nicotinic acid** (a form of **niacin**, a vitamin) and **clofibrate (Atromid-S)**. (Two other drugs, dextrothyroxine and estrogen, were dropped from this study because of serious side effects.)

This study involved men who already had coronary heart disease. Half were treated with either of these cholesterol-lowering drugs; half were treated with a placebo. The purpose was to see if heart disease was less likely to recur in the men treated with the drugs.

In comparison to the placebo group, the incidence of coronary heart disease was almost 20 percent less in men taking nicotinic acid; and it was 9.5 percent less in the men taking clofibrate. Even in these men, for each 1 percent decrease in blood cholesterol there was about a 2 percent decrease in recurrence of coronary heart disease, the same reduction in risk for heart disease as other studies found for people with no apparent signs of heart disease.

**Q:** Any other studies I should know about?

**A:** Yes. The Stockholm Ischemic Heart Disease Secondary Prevention Study, which ended in 1988, involved heart-attack survivors who took a combination of either two cholesterol-lowering drugs, clofibrate and nicotinic acid, or clofibrate and a placebo for five years. At the end of that time, the people taking the two drugs had a 13 percent drop in cholesterol and a 36 percent reduction in deaths from coronary-artery disease.

**Q:** Anything newer than that?

**A:** The latest study, the Scandinavian Simvastatin Survival Study ("4-S" for short), was conducted at 94 medical centers in Denmark, Finland, Iceland,

Norway and Sweden. It found that people previously diagnosed with atherosclerosis who took the cholesterol-lowering drug **simvastatin (Zocor)** were only about half as likely to die of a heart attack as people taking a placebo. In addition, chances of having a nonfatal heart attack or of requiring surgery to open or bypass blocked arteries were cut by about one-third in people taking simvastatin. Incidentally, both groups also ate a reduced-fat diet.

This study, completed in 1994, involved 4,444 men and women ages 35 to 70, with cholesterol levels of 212 to 309 mg/dL. They were followed an average of about 5½ years.

Many doctors consider this an important study because it is the first cholesterol-lowering study to show a reduction in *overall* death rates, not just deaths from heart attacks. People taking the drug were about one-third less likely to die prematurely of a cause other than heart disease.

## LOWER CHOLESTEROL MIGHT NOT MEAN LONGER LIFE

**Q:** **Are you saying that the other studies before this didn't find that lowering cholesterol helped people live longer?**

**A:** Yes, that's exactly what we're saying. Even though lowering cholesterol helped to reduce deaths from heart disease and related problems, a number of studies have found that men with very low cholesterol—usually 160 mg/dL or below—were more likely than men with normal cholesterol levels to die of causes other than heart disease.

**Q:** What were these men dying of?

**A:** A variety of causes, including accidents, suicides and gallbladder and liver problems.

**Q:** Well, we've all got to die of something, right?

**A:** Yes, but that doesn't explain this finding. "The deaths were premature. And they are not easily explained," says David Jacobs, Ph.D., of the division of epidemiology, University of Minnesota School of Public Health, Minneapolis. Jacobs and his colleagues coordinated a study that reviewed data from 20 cholesterol studies from around the world, looking at the link between low cholesterol and death.

**Q:** Do researchers think that low cholesterol *caused* these deaths?

**A:** No one knows for sure, Jacobs says. One class of cholesterol-lowering drugs, **fibric acid derivatives**—clofibrate and gemfibrozil—may have contributed to the deaths. (We talk more about these drugs in Chapter 6.) But researchers are also looking into the possibility that low cholesterol levels may change hormone levels in a way that contributes to health problems, including depression and compulsive behavior. That's something we look at in more detail in Chapter 4.

One other thing you should know about: Statistical findings of studies are sometimes cited in a way that makes the potential benefit of cholesterol-lowering appear greater than it really is in some cases, especially for people who don't have heart disease.

## RELATIVE, ABSOLUTE AND ATTRIBUTABLE RISK

**Q:** What do you mean?

**A:** Take the Helsinki Heart Study we just mentioned, for instance, although you could do the same thing with any study. In that study, the researchers concluded that five years of treatment with a cholesterol-lowering drug resulted in a 34 percent reduction in the risk of nonfatal heart attacks and death caused by coronary-artery disease.

**Q:** Well, that seems straightforward enough. What's the problem?

**A:** Bear with us a moment and look at the actual numbers. In the control group of 2,030 people, there were 84 coronary-heart-disease events (a rate of 4.1 percent); in the treatment group of 2,050 people, there were 56 coronary-heart-disease events (a rate of 2.7 percent).

**Q:** So what's the problem?

**A:** Well, the reduction in actual risk, or **absolute risk**, as it's called in statistics, is only 1.4 percent (4.1 minus 2.7). That means that if 100 people took a cholesterol-lowering drug for five years, only 1.4 coronary events, such as heart attacks, would be prevented. It means that 98.6 people would be taking a cholesterol-lowering drug, exposing themselves to the risks associated with taking that drug, without seeing any benefit.

**Q:** But what about the 34 percent reduction?

**A:** That number is the **relative risk** reduction, a comparison of the risks between two different groups. In this case, researchers take the difference in risk between the control group and treatment group (1.4 percent) and divide it by the risk in the control group (4.1 percent). While that number may be relevant to public-health officials trying to figure out how many fewer people will develop a particular disease as a result of some sort of intervention, it's not especially helpful to individuals who are trying to figure out the best way to allocate their limited time, money and energy in terms of good health.

# Q: I think I know what you mean, but could you say it again?

# A: People like to know their odds, or chances of something happening, before they make big decisions and take risks or spend money. They also like to know how those odds can be changed.

For instance, let's say you're told your risk of having a heart attack in the next five years is 5 in 10. (That's the total absolute risk.) You're also told that if you stop smoking, your odds will drop to 3 in 10; that if you also get your cholesterol down to 200 mg/dL, your odds will be 2 in 10; and if you can, in addition to the above things, get your blood pressure down to normal, your odds will be 1 in 10. You can see exactly how each one of these changes in behavior affects your actual, or absolute risk.

# Q: Do researchers have any other special names for types of risk?

# A: The amount of risk that can be specifically pinned on a risk factor, such as smoking, re-searchers call the **attributable risk**. In the previous example, the attributable risk reduction of smoking is two: the absolute risk, five, minus the risk if you stop smoking, three. The attributable risk reduction of both cholesterol-lowering and blood pressure control is one each.

**Q:** Is knowing the attributable risk important?

**A:** Yes. It can help you figure out which health risk factors are boosting your total risk the most and, therefore, which are most important for you to change. It lets you know which interventions are going to give you the most payback—reduction in risk—for your efforts.

By measuring relative and absolute risk in large groups of people, studies such as the Helsinki ultimately help doctors weigh each individual's attributable and absolute risks.

**Q:** How are you supposed to determine your absolute risk for something?

**A:** "Ask your doctor about this," recommends Donald Vine, M.D., an associate professor of medicine at the University of Kansas School of Medicine, Wichita. "When your doctor prescribes some treatment, ask questions to find out exactly how much a difference it can really make for you. If it's a cholesterol-lowering drug that's being prescribed, ask questions like 'What are my chances of having a heart attack if I don't take this drug? What are my chances if I do decide to take this drug?'

"Get your doctor to tell you what your real benefit is, in terms that you can understand," Vine says. "That way, you have the information you need to decide for yourself if that is what you want to spend for that benefit."

# 3 CHOLESTEROL TESTING

## GUIDELINES FOR CHOLESTEROL TESTING

**Q:** Who should have their cholesterol measured?

**A:** According to the National Cholesterol Education Program (NCEP), a group that includes medical organizations such as the American Medical Association and American College of Cardiology, anyone age 20 or older should have his or her total cholesterol and HDL cholesterol measured at least once every five years.

**Q:** How about kids? Do they need to have their cholesterol checked?

**A:** Pediatricians normally don't check a child's cholesterol until at least age two. After that age, NCEP recommends testing only in children with a family history of premature heart disease (heart attack before age 55) or other apparent risk factors, such as severe obesity. One reason you might have a child's cholesterol checked this early is to detect rare but serious inherited cholesterol disorders that can lead to problems, including heart disease, early in life. See Chapter 4 for more on children and inherited cholesterol disorders.

## Q: And teenagers—when should they be tested?

A: A teenager's cholesterol levels might be checked any time if he has never been checked previously, especially if the teen has a family history of premature heart disease or other risk factors, such as smoking, obesity or high blood pressure.

## Q: If I already know I have high cholesterol, should I have it checked more often?

A: Guidelines vary as to when you should have your cholesterol rechecked. It depends on your levels of different lipoproteins, the number of risk factors you have for heart disease and how aggressively you're trying to lower your cholesterol. Your best bet is to ask your doctor if she is following NCEP guidelines for cholesterol testing, which help doctors make complicated decisions regarding diagnosis and treatment of high cholesterol levels.

If your doctor is not following NCEP guidelines for your case, you should know why. Does she think her own way of doing things is better? Or is it that she simply isn't aware that the guidelines exist?

**Q:** What exactly does a cholesterol test measure?

**A:** If you're getting tested at a mall or doctor's office where a few drops of blood are taken from your fingertip, the test can measure total cholesterol and sometimes HDL cholesterol.

If tubes of blood are taken from a vein in your arm (called venous blood), more things can be measured in the blood—total cholesterol, HDL and triglycerides. (LDL cholesterol is measured indirectly, by subtracting other components from your total cholesterol.)

**Q:** If they're measuring all those things, is that still called a cholesterol test?

**A:** Your doctor might call it that, but technically this test is called a **complete lipid profile**.

**Q:** Is that the kind of blood test I'd have to fast for?

**A:** Yes. It's possible to get fairly accurate measurements of total cholesterol and HDL cholesterol even if you've eaten recently, since these lipids don't change much with a meal. But triglycerides and LDL cholesterol are influenced by what you've just eaten. Their levels start to rise within an hour or two of a meal, and stay elevated for 12 hours or so.

To eliminate this as an influence on your cholesterol reading, you'll be told to fast (go without eating) for about 12 hours. Usually you are told not to eat anything after 8 p.m. at night. Then the blood is drawn first thing in the morning.

**Q:** What happens if I *do* eat something?

**A:** Even having as little as a cup of coffee with milk can throw off your results, so you're better off abiding by the no-food rule or, if you forget and have something to eat, rescheduling your test for another day.

## PROBLEMS WITH ACCURACY

**Q:** I've heard that those finger-prick tests offered at malls and health fairs aren't very accurate. Are they?

**A:** "No one know exactly how accurate they are. There are simply no data available to testify to the accuracy of the tests performed in malls," says John W. Ross, M.D., chairman of the College of American Pathologists' Chemistry Resource Committee.

There are licensing regulations for commercial medical laboratories, but regulations to supervise testing at health fairs are much less rigorous and are spotty, depending on the state, Ross says. So, in effect, you are taking your chances that the person doing the test is properly trained and that equipment being used is properly calibrated.

**Q:** Does that mean I shouldn't bother to have this sort of test done?

**A:** That's up to you, but you shouldn't consider the results reliable. Whether the results are high, low or in between, you shouldn't rely on them as the only measurement of your cholesterol.

**Q:** What about these at-home cholesterol testing kits? Are they any more accurate than mall testing?

**A:** These tests, which first came on the market in 1993, sell for $15 to $20 and give you results in about 15 minutes.

One of the more widely distributed brands, Johnson & Johnson's Advanced Care Cholesterol Test, is "97 percent accurate," its manufacturer says. But, again, Ross says, there are no numbers available to reliably determine these tests' accuracy. "The interpretation and validity of the at-home tests are open to serious question," he says. So don't rely on them as your only source of information about your cholesterol.

**Q:** Why would someone use one of these tests?

**A:** Good question. Unlike the home tests used by diabetic people for monitoring blood sugar, cholesterol tests usually do not need to be done quickly or often.

People might use one to screen themselves for high cholesterol, just as might be done at a health fair. Or if they're trying to reduce their cholesterol with diet, weight loss or drugs, they might use it between doctor visits just to satisfy their own curiosity as to how well they're doing.

**Q:** These tests use so little blood. How do they work?

**A:** The blood reacts with chemical-impregnated filter paper. A chemical in the paper converts cholesterol to hydrogen peroxide. The more cholesterol, the more hydrogen peroxide produced. The filter paper then wicks the hydrogen peroxide up, and your cholesterol "reading" depends on how far along the paper the hydrogen peroxide has been drawn. A scale alongside the paper gives your reading. You can put too much blood into one of these tests and still get an accurate reading, since excess blood is squeezed off. But too little blood will give you a falsely low reading.

**Q:** Is a doctor's office cholesterol testing any better than health-fair or at-home tests?

**A:** It's possible that you can get a more accurate reading at a doctor's office than at a health fair. But again, there is no way to know that for sure, Ross says. "Data on this are not available."

**Q:** **What do you mean, there's no way to know for sure? Aren't doctor's office laboratories licensed or regulated?**

**A:** Yes, they are now required to have a federal license and are sometimes required to have a state license. To be licensed, since 1992, they must comply with requirements set by the Clinical Laboratory Improvement Amendments (CLIA).

These requirements were established by Congress specifically to improve doctor's office laboratory testing. Since 1994, they require proficiency testing to prove that the equipment works properly and that the doctor and staff know how to use it, documentation to show that the staff can keep track of each person's blood sample throughout the testing process, and the adoption of quality-control procedures.

However, there has never been much manpower to enforce CLIA requirements, government officials admit, and physician groups are now actively soliciting Congress to remove these requirements.

Another organization, the Commission on Office Laboratory Accreditation (COLA), offers special, voluntary accreditation, which also helps doctors get their laboratories up to par.

However, neither CLIA nor COLA actually monitors the accuracy of testing in doctors' office laboratories, Ross explains. "That's something that would be almost impossible to do, given the wide variety of instruments and techniques used in doctors' offices to measure cholesterol."

**Q:** What's that mean? That test results can be way off without someone knowing it?

**A:** Let's just say that, here again, this is not the most reliable way to have your cholesterol tested.

In fact, a 1995 report from the federal General Accounting Office found that the tabletop analyzers used to measure cholesterol in about 19 percent of doctors' offices and at most health-fair cholesterol screenings have an error rate of 17 to 50 percent.

"It's true that there are desktop analyzers that give inaccurate results in some studies, but most desktop analyzers are capable of giving good results," says James Cleeman, M.D., coordinator of the National Cholesterol Education Program. "Still, it's perfectly reasonable to ask your doctor if he does use a desktop analyzer, and if so, what he does to ensure that it gives him accurate results."

Some doctors' offices have their own small clinical laboratories, and instead of using tabletop analyzers use smaller versions of the same sort of analyzers used at large professional labs. But here again, you can ask: "What do you do to ensure the accuracy of your results?" Cleeman says. "Ideally, you'd like to hear that the doctor participates in a program of external standardization that allows him to measure his results against results that are known to be accurate."

**Q:** What if my doctor sends my blood to a lab for analysis? Is that the best way to get accurate results?

**A:** Most experts agree it is.

"The odds are very good that you'll get accurate results if the test is done in a licensed laboratory, either in a medium-size to large hospital or a laboratory that does interstate business," according to Ross.

Some 6,000 of these laboratories participate in a special survey conducted every few years by the College of American Pathologists. The survey is designed to assess several important aspects regarding a laboratory's ability to consistently come up with the right results, Ross says. The last survey, done in 1994, showed that 97.5 percent of the laboratories met the NCEP Criteria, which require that individual cholesterol results be off target by no more than plus or minus 8.9 percent.

## Q: How can I make sure the lab my doctor uses is this accurate?

A: Ask your doctor if the laboratory she uses participates in the College of American Pathologists' survey, Ross suggests.

If it doesn't, ask your doctor if she knows the lab's credentials. Is it a large hospital or interstate lab? Does it have special credentials in addition to a federal and state license, such as the approval of the College of American Pathologists' Commission on Laboratory Accreditation? You want to know why your doctor uses this particular lab and be assured that its results are checked regularly for accuracy.

## Q: What other kinds of things can interfere with a cholesterol test's accuracy?

A: All sorts of things, especially *you,* can have an impact on your cholesterol test. "The main source of misinterpretation of a cholesterol test is not the laboratory, it is the person," Ross says.

**Q:** What causes this variation in cholesterol levels in people?

**A:** Your body can. A reading may change depending on whether blood is drawn from a capillary or a vein, whether you are standing or sitting, the composition of recent meals, whether you've been ill recently, if you're pregnant or taking drugs, whether you are losing or gaining weight, even the time of year. (Cholesterol tends to be slightly lower during the summer months.)

## THE BEST WAY TO TEST

**Q:** So what's all this mean? I shouldn't even bother to have my cholesterol tested?

**A:** No, but have it done right. Here's what Ross suggests:

• Make sure all elements of your health status—weight, diet, medications—have been stable for at least four months.

• Have the test performed on a venous sample (blood drawn from a vein, usually in your arm), not capillary blood (from a finger prick).

• Limit your physical activity to quiet walking before the test, and sit quietly for 10 to 15 minutes before the sample is drawn.

• Don't allow a tourniquet to remain on your arm for more than 30 to 60 seconds before drawing blood.

- Have your sample sent to a major commercial laboratory or a large hospital laboratory for analysis.

**Q:** **So if I do all those things, am I assured of accurate results?**

**A:** You're as close as you're going to get.

**Q:** **So is that all I need to do?**

**A:** Possibly. If you follow the above advice and your total cholesterol reading is lower than 180 mg/dL, "you can be fairly well assured that you are not at increased risk for heart disease and should have no further testing at the time," says Ross. Your HDL should also be higher than 35 mg/dL.

**Q:** **What if my cholesterol reading is higher than that?**

**A:** If your total cholesterol is above 180 mg/dL, there is enough possible variation that you don't really know what risk group you are in, according to Ross. Your average, day-to-day value could be higher than 200 mg/dL.

So you'll want to wait three or four weeks (to help minimize the impact of seasonal variations in cholesterol), then repeat the test at the same lab, in the same manner.

## Q: What's that supposed to do?

A: You can compare the two results to see how close they are. If the first two readings are within 15 percent of each other, you can average the two and use that number as your guide. In other words, if one reading is 210 mg/dL and another is 240 mg/dL, add them together, divide by two, and come up with 225 mg/dL as your actual cholesterol level.

## Q: What if the two readings are way off?

A: If the second result is more than 15 percent different from the first result, review your health status and pretest preparation period. Were there any changes? If not, get a third reading, again waiting three to four weeks, Ross advises. Don't switch laboratories during a series of readings, however. Or if you do, start the series over. "You don't want to throw in yet another variable—a different lab," he says.

## Q: Now, what do I do with the three numbers?

A: Throw out the one that's farthest from the other two (the outlier) and average the two that are within 15 percent of each other. If none is within 15 percent and the lab is a good one, either your health status, the pretest preparation or the blood-drawing and sample-handling processes are not stable. "This is an unusual circumstance," according to Ross, "so step

back and think about the variations that could be
affecting your results before beginning again.''

**Q:** **If my initial cholesterol reading is
high, should I also go through this
additional testing?**

**A:** If your total cholesterol is 240 mg/dL or higher,
you should get a second reading, but ''don't
wait to get a second test to begin to do something
about it,'' Ross says. ''My thinking is that people
shouldn't need the inducement of a high cholesterol
reading to lose weight, stop smoking and start exercising.'' However, any decision to start taking
cholesterol-lowering drugs should be based on
more than one measurement.

**Q:** **What about checking my HDL and LDL
levels? Is that important?**

**A:** Yes. Most doctors believe it's important for
anyone with a total cholesterol reading of
200 mg/dL or higher (some doctors even say as low
as 180 mg/dL) to have his HDL, LDL and triglyceride
levels measured. In fact, the National Institutes of
Health has now recommended that HDL be measured
as part of the initial cholesterol assessment in
all adults.

**Q:** **Why are these other measurements so important?**

**A:** They take into account the breakdown of cholesterol in your blood, not just the total amount. In other words, they seem to give a more accurate assessment of your real risk for developing heart disease than does total cholesterol alone, experts say.

**Q:** **How so?**

**A:** The fact is that most of the people who have heart attacks don't have superhigh cholesterol readings. They have cholesterol levels between 200 and 250 mg/dL, just like most people. So the trick is to identify the people in this group at highest risk for heart attacks.

**Q:** **How is that done?**

**A:** All of a person's risk factors need to be weighed —whether she's overweight, diabetic, smokes or has a strong family history of heart disease, for instance.

And in addition to that, doctors want to know a person's levels of HDL and LDL cholesterol and, usually, triglycerides, because each of these blood components is considered to carry its own risks for heart disease.

# HDL AND LDL LEVELS AND RATIOS

## Q: What's considered a good level of HDL cholesterol?

**A:** Anything below 35 mg/dL has been found to increase a person's risk for heart disease, although some people with very low total cholesterol (160 mg/dL or lower) have an HDL level below 35 mg/dL with no apparent increased risk. Levels of 35 to 50 or so are considered normal, but some people's HDL can reach 75 mg/dL or higher. In several large studies, a 1 mg/dL increase in HDL cholesterol was associated with a 2 to 3 percent decrease in risk for coronary heart disease.

## Q: And what levels of LDL cholesterol are good?

**A:** Amounts less than 130 mg/dL are considered desirable, and amounts less than 100 mg/dL are associated with a reversal of coronary heart disease. Borderline high levels are 130 to 149 mg/dL; anything over 150 mg/dL is high.

## Q: How do doctors juggle all these numbers?

**A:** One thing they sometimes do is come up with ratios of one blood lipid to another. Researchers, especially, are looking at ratios because they appear to

be better predictors of heart disease than is total
cholesterol alone.

# Q: What kind of ratios do they use?

A: Total-cholesterol-to-HDL is one ratio that
seems to help pinpoint risks. In studies, a total-
cholesterol-to-HDL ratio of less than 6:1 seems to
protect against heart disease, while a ratio greater than
6:1 is predisposing.

Some doctors also determine LDL-to-HDL ratios. In
studies, an LDL-to-HDL ratio of less than 4:1 is protective, whereas a ratio higher than 4:1 is predisposing to
heart disease.

# Q: My doctor never mentioned ratios to me. Can I figure this out on my own?

A: Yes, if you know your total cholesterol, HDL
and LDL. Your total-cholesterol-to-HDL ratio is
simply total cholesterol divided by HDL. Same goes
for LDL-to-HDL. Simply divide the first figure by
the second.

**Q:** Does any ratio stand out as being a particularly good predictor of coronary-artery disease?

**A:** So far, total-cholesterol-to-HDL seems to be the best predictor. But experts disagree on the value of ratios, and some say ratios can be misleading and that you should always look at the numbers themselves, too.

## TRIGLYCERIDE LEVELS

**Q:** What about triglycerides? When should they be measured?

**A:** Doctors say you should have your triglycerides measured if you find you have total cholesterol higher than 200 mg/dL, if you have two or more coronary-heart-disease risk factors, diabetes, high blood pressure, kidney disease or a condition called **pancreatitis** (inflammation of the pancreas).

**Q:** What are considered good levels of triglycerides?

**A:** Normal levels are less than 200 mg/dL; borderline high, 200 to 400 mg/dL; high, 400 to 1,000 mg/dL and very high, 1,000 mg/dL or above.

**Q**: Enough already with the numbers.
What's this mean to me, the consumer?

**A**: The point is this: Your doctor is going to use the
results of your cholesterol tests to help make
decisions regarding your medical treatment. You want
to make sure he is working with accurate numbers and
that he knows how to interpret those numbers. So ask
for a copy of your cholesterol test. Know your total
cholesterol, your HDL, LDL and triglycerides levels.
Check the table on page 59 to see what risk categories
you fall into. And check the list on page 60 to see what
other risk factors for coronary heart disease you have.

Ask your doctor to explain his treatment decisions,
and what they are based on.

And remember, high cholesterol by itself is not a
"disease." It's simply one among many risk factors for
heart disease.

## TABLE A

---

## National Cholesterol Education Program Cholesterol Classification Guidelines for Adults

---

► **Total cholesterol**

Desirable: Less than 200 mg/dL
Borderline high: 200-239 mg/dL
High: 240 mg/dL or higher

► **HDL cholesterol**

Desirable: 50-75 mg/dL or higher
Borderline low: 35-49 mg/dL
Low: Less than 35 mg/dL

► **LDL cholesterol**

Desirable: Less than 130 mg/dL
Borderline high: 130-159 mg/dL
High: 160 mg/dL or higher

► **Triglycerides**

Safe: 200 mg/dL or less
Borderline: 200 to 400 mg/dL
High: 400 to 1,000 mg/dL
Very high: 1,000 mg/dL or above

---

## TABLE B

---

## Additional Risk Factors for Coronary Heart Disease

---

High cholesterol is only one risk factor for coronary-artery disease. These additional risk factors are just as important:

- ► Age 45 or older in men
- ► Age 55 or older in women
- ► Cigarette smoking
- ► Diabetes mellitus (noninsulin-dependent, or adult-onset)
- ► HDL less than 35 mg/dl
- ► History of **cerebrovascular** (stroke) or **occlusive peripheral vascular disease** (intermittent claudication)
- ► High blood pressure
- ► LDL 130 mg/dl or higher
- ► Heart attack or sudden death before age 55 in a parent or sibling
- ► Severe obesity (30 percent or more over ideal weight)
- ► Total cholesterol 200 mg/dl or more

---

# 4 SPECIAL INTEREST GROUPS

## WOMEN AND CHOLESTEROL

**Q:** I just had my cholesterol checked and it was 250 mg/dL. The doctor said it was okay because I am a woman. What does she mean?

**A:** Your doctor may be referring to results from the ongoing, two-decade-long Framingham Heart study, one of the few cholesterol-monitoring studies to include women. Researchers found that what's considered a high total cholesterol level for men—and therefore an important risk indicator for heart disease —does not correlate to women. In general, women do not begin to have heart-disease-related problems, such as heart attacks, until their cholesterol levels reach 265 mg/dL. That's 25 above the danger zone for men, which begins at 240 mg/dL.

**Q:** Does this mean women can tolerate higher levels of cholesterol and can ease up about buttering toast?

**A:** Sorry, there's no indication that women are free to binge on butter or other artery-clogging foods without suffering cholesterol consequences. What seems clear is that women normally have higher overall cholesterol levels than men because of the positive effects of the female hormone estrogen, not because they can tolerate more plaque-building cholesterol in their arteries.

**Q:** Why? What's estrogen got that's so helpful?

**A:** Estrogen boosts the production of HDL, the cholesterol that escorts fat from the bloodstream and slows plaque buildup in the artery walls, which is caused by LDL. Thanks to estrogen, women, on average, have an HDL level of 55 mg/dL compared with an average of 45 mg/dL for men. In short, their estrogen edge gives women built-in heart-disease protection— at least until menopause.

**Q:** Then what happens?

**A:** Once a woman reaches menopause and estrogen levels naturally decline, HDL cholesterol production drops. And there goes the natural protection against plaque buildup. The Framingham folks found that if HDL levels dip—even a little—a woman's heart-disease risk rises greatly. And in the Lipid Research Clinic's Follow-up Study, women who had as little as a 10 mg/dL drop in HDL doubled their risk of heart disease. One proven method women can use to compensate for this drop is to take estrogen-replacement hormones, a course known as hormone-replacement therapy (HRT).

**Q:** **You mean I could have low total cholesterol but still be at risk for heart disease because my HDL level is also low?**

**A:** Absolutely. To determine if that 250 cholesterol reading of yours is really okay, you need to know your HDL level, which provides a more precise indicator of heart-disease risk than total cholesterol alone. But you can't ignore your LDL either. While these "bad guy" lipids do not seem to be as important predictors of heart disease in women as they are in men, "LDLs are still the culprits that lay down the plaque on arteries and cannot be completely wiped out by HDLs," says Valery T. Miller, M.D., professor of medicine and director of the Lipid Research Clinic at George Washington University Medical Center, in Washington, D.C.

**Q:** **Okay, so how high should my HDLs be? How low should my LDLs be?**

**A:** As a general rule of thumb, women should have HDLs higher than 55 mg/dL (the NCEP recommends that HDLs be higher than 35 for both sexes); and LDLs should be lower than 130 mg/dL. With that said, however, the Framingham study indicates that it's more precise to know the *ratio* of total cholesterol to HDLs in your blood. As we explained in Chapter 3, ideally the ratio of total cholesterol to HDL levels should not exceed 6:1. Above that and your heart-disease risk soars. Furthermore, your LDL-to-HDL ratio should not exceed 4:1. These same safe ratios, by the way, also apply to men.

**Q:** **I'm having a math attack. Can you review this ratio again?**

**A:** If your total cholesterol reading is 250, for example, then your HDL should be 50 or more to stay within the recommended 6:1 safe zone. If your LDL is 140, then your HDL should be 35 or higher so as not to top the 4:1 safe zone.

**Q:** **Okay, so I should know my HDLs and LDLs and not worry about the total cholesterol?**

**A:** Not exactly. The Framingham study showed that women under age 50 who had the highest total-cholesterol levels were most likely to develop heart disease.

Until there's more conclusive data on women and heart disease, most experts advise you to keep your total cholesterol at 200 mg/dL or less, no matter what your sex. This involves reducing the amount of high-fat foods in your diet, exercising and controlling other risk factors, such as weight gain and smoking. Smoking, by the way, reduces estrogen, lowers HDLs and raises LDLs, prompts the onset of early menopause, and shifts

body-fat distribution to your waistline. This fat shift
creates an "apple shape," which is associated with
higher LDLs.

**Q:** **My doctor told me that triglyceride levels really determine a woman's heart-disease risk. Is that true?**

**A:** Triglycerides are blood fats that come from the diet or are made in the liver. And it's true, their link with heart disease is stronger in women than it is in men. While the role of triglycerides is still uncertain, their significance to women's heart-disease risk appears to be a case of "guilt by association" to HDLs. Both lipids are affected by estrogen. Studies have shown that after menopause, when estrogen wanes, HDLs nosedive while triglycerides rise, which is bad news. Studies also show that when women take synthetic estrogen, HDL levels rise, but so do triglycerides. Some researchers believe that synthetic estrogen may alter HDL, making it less able to remove cholesterol. (By the way, hormone-replacement therapy that also includes **progesterone** appears to raise triglycerides less significantly.)

While studies show a connection between HDLs and triglycerides, the connection to developing heart disease remains unclear. Some experts say that high triglycerides is a marker, or an indicator, that you may have an underlying problem in processing blood fats, possibly due to estrogen preparations or other medications or diabetes, that puts you at increased risk of coronary heart disease.

Indeed, the Framingham study shows that high triglyceride levels (above 190 mg/dL) are premier predictors (or markers) for heart disease in women, particularly postmenopausal women, but not for men. In younger women, the estrogen hormone may override any potential harm or produce a more benign form of triglycerides, according to Miller. In any case, to get a true read on your cholesterol as well as your risk, get a complete lipid workup that includes triglyceride levels.

**Q:** At what age should women be tested for cholesterol?

**A:** Some doctors suggest that, in general, women with a family history of heart disease should have a lipid profile test done before age 18 and be closely monitored thereafter. Women without a family history of heart disease should have a complete cholesterol workup at age 30 and certainly before menopause, when lipids change.

**Q:** Before prescribing the Pill, my doctor wants my cholesterol checked. Why is that?

**A:** The Pill has not always been shown to be lipid friendly. Studies done on earlier, more potent versions of the Pill, which contained high estrogen doses, showed that Pill users had a threefold increase in death from heart disease compared with non-Pill users. Studies have also shown that oral contraceptives that contain progesterone can cause blood clots and may also change fat and carbohydrate metabolism. The Pill is a good news/bad news story. The high-dose estrogen component boosted triglycerides and, more significantly, the good guy HDLs—but unfortunately, not enough to compensate for the raised LDLs, which got bumped up, thanks to the high-dose progesterone component.

## Q: What about newer versions of the Pill?

## A: The newer, low-dose Pills, which contain less-potent estrogen and progesterone, appear to only mildly increase LDLs while increasing HDLs. But you still need a complete cholesterol workup before going on any oral contraceptive. An annual checkup is also necessary. Some doctors rule out the Pill if cholesterol levels exceed 220 mg/dL or triglycerides exceed 190 mg/dL. Other doctors believe that elevated cholesterol or triglyceride levels should not necessarily rule out Pill use, "since the estrogen component would cancel out harmful LDLs and triglycerides," Miller says. However, oral contraceptives are taboo if you are extremely overweight, smoke or have very high blood pressure.

## Q: Let's say I'm menopausal and, since heart disease runs in my family, my doctor recommends hormone therapy. Is that because it lowers cholesterol?

## A: Replacement estrogen affects cholesterol in two ways: It increases HDL and decreases LDL, both by 10 to 15 percent, according to some studies.

Postmenopausal hormone-replacement therapy has been shown to cut women's risk of heart disease in half. But it's not for everyone and it's not the only

weapon in the cholesterol-busting arsenal. Talk with
your doctor about diet, exercise and weight control.
Consult books that address "the HRT decision," review
all your risk factors and get a second opinion.

**Q: I've heard that the newer version of HRT reduces uterine cancer but may cancel out heart protection and also increase breast-cancer risk. True or false?**

**A:** Recent results from the National Institute of
Health's Postmenopausal Estrogen/Progesterone
Intervention (PEPI) study may cast HRT in a new, more
favorable light. PEPI shows that estrogen used alone or
in combination with a progestin (synthetic progesterone,
added to counter harmful uterine-tissue buildup) does
not, in fact, block estrogen's positive effects on lipids.
The HRT combo also improves HDL levels, although
the increase is only slight. If estrogen alone is used,
there is a 5 mg/dL rise in HDLs; if progestin is added,
HDLs rise less than half that amount.

Experts predict that updated HRT formulas with a
micronized, gentler form of progestin (similar to the
type in the Pill) will soon be available. These formulas
will have even less negative impact on the estrogen's
positive effects. In the meantime, experts point out
HRT's other heart-healthy benefits: It lowers high
blood pressure and protects the arteries. The risk of
breast cancer does increase, and that has had the lion's
share of publicity regarding HRT. But breast cancer
causes far fewer deaths than heart disease, and, as
Miller states: "The benefit of sparing thousands of
women early death from heart disease outweighs the
possible risk of breast cancer so convincingly that the

only women I would exclude from HRT use are those with a blood clot or liver disease.''

**Q:** **Which form of HRT protects better: the patch or pill?**

**A:** The Pill—oral contraceptives—lowers cholesterol best, because the hormone must pass through the liver to be metabolized, so it has a direct effect on liver function. (The liver, you'll recall, makes cholesterol for the rest of the body.)

A skin patch, on the other hand, provides estrogen directly through the skin and, from there, into the bloodstream. Although some estrogen does reach the liver, its effects are much less. Still, women who use the patch do gain some protection from heart disease. Indeed, the patch may be the preferred method of HRT if you have very high triglyceride levels, since oral estrogen may raise triglycerides.

## CHILDREN AND CHOLESTEROL

**Q:** **Our family doctor says it's never too early to start a heart-healthy diet. Just how early does heart disease begin anyway? Are my kids at risk?**

**A:** Some experts say it's overkill to start kids on a low-fat diet unless they're overweight or have extremely high cholesterol. But there is some evidence that as early as age three arteries may begin clogging with fatty streaks made up of fat and connective tissue. Apparently, too, the more extensive the streaks in childhood the more likely the clogging is to advance to atherosclerosis.

A few years back, researchers in the Bogalusa (Louisiana) Heart Study looked at cholesterol levels of children for 15 years. Autopsies of the children in the study who died accidentally during this time revealed that those who had higher levels of overall cholesterol in earlier years had more extensive fatty streaks in their aortas. While there's no definite proof, this and other evidence leads some experts to believe that children with high cholesterol levels are more likely than the general population to have high cholesterol as adults.

**Q:** **What do fatty streaks have to do with what my kid eats?**

**A:** As we explain in Chapter 5, fatty streaks are caused by high blood levels of cholesterol, which usually are a result of eating foods high in saturated fats. Preference for this sort of food—burgers, French fries and the like—is determined early in life. Unfortunately, the earlier your child starts eating high-fat foods and the more he eats, the more likely it is that he will develop the high cholesterol levels that put him at risk for heart disease. A high-fat diet is a particularly important risk factor if premature heart disease runs in the family—before age 55 in men and before age 65 in women.

**Q:** So how do I know if my child's cholesterol is high?

**A:** Most doctors don't routinely screen children for high cholesterol unless one or both parents have high cholesterol. So if either parent has a total cholesterol level above 240 mg/dL, and further tests reveal HDLs lower than 35 mg/dL, LDLs higher than 130 mg/dL or triglycerides above 250 mg/dL, it's time to schedule an initial measurement of total cholesterol for your child. High cholesterol can be passed on from parent to child, and the earlier you identify it and take measures to control it, the less likely it is to progress.

**Q:** But my kid is not even school age. Isn't that a bit young for screening tests? What do the tests involve?

**A:** He's not too young if there's a family history and a parent's blood levels indicate the child could be at high risk. Then any time after age two is acceptable to have a child tested for cholesterol, and those children whose parents have high cholesterol should certainly have testing before age 10, some experts contend. Some doctors also recommend cholesterol testing for children who are excessively overweight or for teens who smoke or use oral contraceptives.

**Q:** Does the doctor have to draw blood from a vein?

**A:** Perhaps not. Your doctor may first give your child a simple finger-prick blood test. Ideally, the level should be under 170 mg/dL for kids ages 2 to 19. If it's below this level and you're gung-ho about it, you may still want to limit the burgers, promote exercise and discourage smoking. In five years, your child should be rechecked for cholesterol levels.

**Q:** And if my child's cholesterol is high?

**A:** There's a one in four chance your child's cholesterol *will* be higher than 170 mg/dL. A high cholesterol reading means that your child needs a further, more detailed test—the complete lipid profile we discussed in Chapter 3. That test involves drawing blood from a vein following a 12-hour fast. A complete lipoprotein analysis is also recommended if either parent or grandparent has documented premature heart disease, such as a heart attack before age 55.

**Q:** It's not so easy keeping kids from eating for 12 hours, let alone getting them to sit still for a needle. How important are these tests?

**A:** Very. It's the only way to accurately assess a child's real risk. A complete lipid profile can tell you if your child's high cholesterol is due to raised HDLs (which is good and affects about 15 to 20 percent of children with high overall cholesterol) or if it's due to raised LDLs, the artery-clogging lipids. This is not so good and occurs in 1 in 20 kids. The test can also reveal if your child's triglyceride levels exceed the recommended level of 100 mg/dL for kids under age 10 or 130 for kids over that age.

**Q: I take it high triglycerides is a bad sign in kids?**

**A:** Yes, as it is in adults. High triglycerides are a tip-off that children may have an inherited condition associated with premature heart disease, which occurs in 5 percent of kids. Basically their bodies are unable to metabolize fat.

**Q: If early high cholesterol levels might indicate premature heart disease, why don't they routinely test for it in school?**

**A:** Experts believe that there is not enough ironclad evidence to show just how well childhood cholesterol levels predict adult levels to warrant universal screening of all kids. In fact, studies show that quite a few children with high cholesterol levels will not have high enough levels as adults to qualify for individualized treatment. The harm of universal screening, experts argue, is that it would single out lots of children for repeat testing and perhaps prompt unwarranted treatment and the overuse of cholesterol-lowering drugs.

**Q:** What studies are you talking about?

**A:** In one study published in the *Journal of the American Medical Association* (December 19, 1990), researchers at the University of California in San Francisco tested the cholesterol levels of 2,367 children, ages 8 to 18, for a period of 10 years. Fifty-seven percent of the girls and 30 percent of the boys had been identified as having levels in the danger zone. But when the researchers looked at lipid tests on these same individuals 20 to 30 years later, only a fourth of the females and less than half of the males who were in the danger zone met the National Cholesterol Education Program's criteria for intervention as adults.

That's why, at this point, the NCEP recommends selectively screening only children who have a family history of premature heart disease or one parent with high cholesterol. Children who are overweight or have other risk factors may also be selectively screened. Of course, selectively screening means that some kids at risk will be overlooked because a parent's cholesterol level or family history is unknown. But in general, most experts agree on this: If your child does not have these risk factors or is not extremely obese or is not eating mainly a high-fat diet, you can relax about cholesterol screening.

**Q:** I have high cholesterol. If my child's test reveals high levels, should I limit pizza and fries and go for low-fat foods?

**A:** You would be on the right track to do so. If a child above age two has an elevated cholesterol level, the first strategy is to change the menu while

maintaining a healthy weight and encouraging physical
activity. A child in the high blood-cholesterol range
who adopts a diet lower in cholesterol and saturated fat
will have lower overall cholesterol and lower LDLs
throughout childhood and very probably lower cholesterol as an adult as well, thus making him less likely to
develop heart disease. In one study, 484 children ages
12 to 18 substituted low-cholesterol/low-saturated-fat
foods for those in the typical American kid's diet and
reduced saturated fat from 15 to 10 percent of their
calories. The results? Their cholesterol levels fell after
less than three weeks on the diet.

As you plan meals, think low-fat rather than diet,
which has come to be equated with deprivation. We talk
more about the benefits of low-fat eating in Chapter 5.

**Q:** **Does a reduced-fat diet provide enough
nutrients for my growing kid?**

**A:** Yes. A child can get the necessary protein, vitamins and minerals for normal growth. In fact,
low-fat diets offer more in the way of nutrients since
they offer lots of fruits, vegetables and grains, along with
low-fat dairy products, lean meat and fish. However,
you might want to consult your physician or nutrition
expert to guide you in preparing a healthy menu.

**Q:** **How long should we give this diet?**

**A:** Give it at least three to six months. If after that
time your child's total and LDL cholesterol
remain above 230 mg/dL and 160 mg/dL respectively,

and she is of normal development and school age, your physician might recommend using a **bile acid binding resin**. This is particularly important if the child has a positive family history of heart disease or a low HDL level.

**Q:** Just what are these resin drugs used for kids?

**A:** The typical bile acid binding resins are cholestyramine or **colestipol**, powders that are mixed with juice and taken during meals, starting with one scoop twice daily and gradually increasing. Long-term studies show these drugs are relatively safe and effective in children, and have fewer side effects than other cholesterol-lowering drugs because they are not absorbed by the intestine. The most common side effects are constipation, nausea and bloating. Close monitoring for growth and vitamin deficiencies is necessary, however, as these drugs may interfere with proper absorption of normal fats and fat-soluble vitamins and folic acid. We discuss these drugs in detail in Chapter 6.

## PEOPLE WITH INHERITED CHOLESTEROL PROBLEMS

**Q:** If my mother or father is being treated for high cholesterol, does this mean I might have the same problem?

**A:** You might. High cholesterol can be due to excess weight and other factors or it can be due to an inherited lipid abnormality. The NCEP reports that among children screened because of premature heart disease in a parent or grandparent, approximately one-third have some form of lipid abnormality.

The most common form of inherited high cholesterol passed by genes is **familial hypercholestemia (FH)**. It's caused by a defect in **LDL receptors**, the portals on membranes that allow cholesterol to move in and out of the cells. When LDL receptors are defective, cholesterol can't be adequately removed from blood, so it backs up and deposits in the arteries.

**Q:** **What are my chances of having FH? And if I do, what can I expect?**

**A:** FH affects 1 in 500 people. If either your mother or father has FH, you have a 50 percent chance of developing heart disease by age 40 if you're male or by age 50 if you're female. People who have inherited FH from both parents, which occurs in one in a million cases, develop heart disease before age 20 and have few, if any, functioning LDL receptors.

**Q:** **Are there any clues to tell who might have FH?**

**A:** The biggest clue is if either parent or grand-parent had a heart attack before age 60. Another clue is if your total cholesterol level is above 300 to 400 mg/dL, or your LDL level is well above 200 mg/dL. People who have inherited the FH gene from both parents have total cholesterol levels that top 600 to 1,000 mg/dL.

**Q:** How about any visible clues of an inherited disorder?

**A:** People with FH may have raised yellowish bumps, called **xanthomas**, which are fatty deposits of cholesterol that settle in the tendons of knees, elbows or knuckles. Xanthomas usually show up in a person's 20s or 30s, but can appear in younger children, who may also have an opaque ring around the cornea of their eyes.

**Q:** What about tests that can show me for sure if I have FH?

**A:** The positive test for FH shows that LDL receptors are defective and involves extracting cells from skin tissue and growing them in a culture. Your doctor may refer you to a lipid specialist who can give you this test. In the near future, tests may be available that help identify genetic blood markers called **apolipoproteins**. These are the protein components of lipids and seem to effect the clearance of cholesterol in people with inherited disorders, according to preliminary results from the Framingham study.

**Q:** I have FH and my daughter (whose cholesterol is normal) is pregnant with my first grandchild. How can we tell if my problem has passed on to the baby?

**A:** In FH babies, LDL levels in blood from the umbilical cord at birth measures twice as high as levels found in normal babies. A blood sample can then be taken after age one. The results will help you get started early at controlling cholesterol so you can retard development of advanced atherosclerosis.

**Q:** My 48-year-old brother survived a heart attack and keeps bugging me to get my triglycerides checked, as I could be next. I'm in my mid-30s. What's he talking about?

**A:** Besides inheriting high levels of cholesterol, people who develop heart disease before age 55 sometimes also inherit high triglyceride levels, which points to a condition known as **familial combined hyperlipidemia (FCH)**. Half of the people who have a first-degree relative—mother, father, sibling—with FCH also develop FCH. So if your HDL is too low, your LDL is too high and your triglyceride level is also high (250 to 500 or above), you probably inherited the FCH gene. And it may have passed to your offspring, so get your children's triglycerides checked.

**Q:** If I have an inherited high-cholesterol problem, am I doomed to a heart attack or stroke?

**A:** No. While you are more likely to have premature heart disease than someone without a family history of FH, you can control your cholesterol and hedge your odds of early heart attack by adhering to a low-cholesterol, low-fat diet. Adopting such a diet helps increase the number of LDL receptors, which in turn leads to a decrease in LDL in the blood. Monitoring your cholesterol will be a lifelong process, however, and you'll also need to control high blood pressure, avoid smoking and shed excess weight so that you're no more than 20 pounds over your ideal weight.

**Q:** When is drug treatment called for?

**A:** That depends on how high your cholesterol level is, how old you are and other risk factors. If your cholesterol is 600 mg/dL and you're 15 years old, it's best to get your cholesterol down as quickly as possible using medication. But if it's 280 mg/dL, it's better to start with dietary measures and wait until age 20 or 25 to start drugs.

For adults with cholesterol levels in the 500 mg/dL range, dietary methods can be started. If these do not work after a fair trial, drugs such as bile acid binding resins will remove LDL cholesterol and lower total cholesterol.

## OLDER PEOPLE AND CHOLESTEROL

**Q:** Let's say I'm over 70. Will there ever be an age when I can stop worrying about cholesterol and eat what I want?

**A:** The verdict is not in yet. However, there is intriguing evidence from a Yale University study (published recently in the *Journal of the American Medical Association*) showing that cholesterol may not be as important a risk factor for older folks as it is for the middle-aged.

For four years, Yale researchers tracked 1,000 people past age 70, with an average age of 79. Their findings: Neither high total cholesterol nor low HDLs appeared to increase the risk of dying of heart attack.

**Q:** **Does this mean that after age 70 I can pass on the cholesterol check?**

**A:** Let's put it this way: According to some reports, there is a push to have elderly people routinely screened for high cholesterol as part of an effort to counter the growing burden of heart disease's consequences. But the Yale researchers believe this is unwarranted. In their view, until there's good evidence to show that benefits of screening and treatment outweigh harm for the elderly, people 65 and older without heart disease should not be routinely screened. Nor should they be treated with drugs, which can have more serious side effects in the elderly. Perhaps their most emphatic stance is that the benefits of cholesterol-lowering treatment based on tests done on middle-aged people cannot—and should not—be extrapolated to elderly people.

**Q:** **Is this no-screening/no-treatment advice for seniors generally accepted among heart experts?**

**A:** Far from it. This stance has been soundly challenged by many leading heart experts, including noted Framingham researcher William Castelli, M.D. These doctors believe that screening and preventive cholesterol-lowering methods can help the elderly avoid debilitating heart attacks and strokes and perhaps, down the road, keep them out of nursing homes. "As people get older, the risk of heart disease increases, which may not be identifiable until an event such as heart attack occurs," says Lewis Kuller, M.D., chair of the department of lipid research at the University of Pittsburgh School of Medicine. "Heart attack and stroke may not kill you but they can leave you less independent and make your health go downhill."

**Q:** So this group says that if I want to hang on to my health I should watch my cholesterol, no matter what my age?

**A:** In Kuller's view, even if you are healthy at age 60, 70 or 80 but indulge in artery-clogging foods, for example, you are courting a catastrophe, such as heart attack or stroke, that could be devastating to your quality of life. On the other hand, evidence is strong that lowering your lipid levels reduces vascular disease. Just how much you need to lower it and by what method, however, depends on the extent of vascular disease. And that can only be revealed by a **carotid ultrasound test**.

**Q:** I smell an expensive test. What does it do exactly?

**A:** It measures the extent of **calcification** (hardening) in the arteries. It's more expensive than a blood test, but compared with the risk of surgery or the possibility of debilitation, it's worth it, Kuller believes. If you have little vascular damage but your cholesterol is elevated, watching what you eat and being active are prudent strategies and work effectively in three-fourths of people, he says. Higher cholesterol or more extensive vascular disease may require cholesterol-lowering drugs, which seem to be well-tolerated by older people, according to Kuller.

**Q:** Does "watching what I eat" mean skipping steak and homemade pie?

**A:** Based on evidence to date, the best advice for older folks may be to take a moderate approach to meals. The Yale researchers believe that overemphasizing a low-fat, low-cholesterol diet when combined with a limited income could disrupt an older person's nutritional intake and interfere with the enjoyment of eating, an important pleasurable activity. So if you aim to eat mainly foods low in saturated fat, you can probably keep the occasional filet mignon, lobster, Boston cream pie and other tasty fare on your menu.

## PEOPLE WITH TOO-LOW CHOLESTEROL

**Q:** You mentioned in Chapter 2 that very low cholesterol has been associated with some health problems. Can you tell me more about that?

**A:** As we said, a number of studies have found that men with very low cholesterol—usually 160 mg/dL or below—were more likely than men with normal cholesterol levels to die of causes other than heart disease.

**Q:** **What about women with low cholesterol? Any potentially negative connection there?**

**A:** Apparently not. These studies involved mostly men, and the link seems to hold only for men.

**Q:** **Were these studies of men with naturally low cholesterol or of people being treated to lower their cholesterol?**

**A:** Both. In the treatment studies, however, only those using cholesterol-lowering drugs, *not* those using diet and lifestyle changes (such as stopping smoking) were more likely to die of non-heart-disease-related illness.

**Q:** **And what were these men dying of?**

**A:** A variety of causes, including some types of cancer, lung or digestive diseases, gallbladder problems, accidents or suicides and, in a few cases, hemorrhagic stroke.

**Q:** How are these illnesses associated with low cholesterol?

**A:** "The truth is, nobody knows for sure what is going on," says David Jacobs, Ph.D., whom, you'll recall, coordinated a review of studies from around the world, looking at the link between low cholesterol and deaths. "We know that these studies find an increased death rate once cholesterol gets below a certain level, but, despite investigation into the possible causes, just why this happens remains murky."

**Q:** Well, how do experts think they might be related?

**A:** There are lots of possible explanations, Jacobs says. One is that cholesterol is lowered by some diseases themselves. People with chronic lung disease often lose weight and so may lower their cholesterol. People with liver disease have reduced production of cholesterol in their livers. And people with digestive problems might not absorb fats very well.

In fact, there's some evidence to show that cancer causes blood-cholesterol levels to drop. Some investi gators have reported that as a tumor grows, the cells actively take up LDL cholesterol, causing blood levels of LDL to drop. And research also shows that people who go into remission as a result of chemotherapy have a return to normal of blood-cholesterol levels.

**Q:** Well, that all makes sense. Why isn't that explanation accepted as the cause?

**A:** Because when these analyses exclude men who died within five or so years after the study ended, which presumably would eliminate most of those who had some kind of disease at the time of the study, the results remain the same: Men with very low cholesterol are still more likely to die prematurely than people with normal levels.

**Q:** What about alcohol abuse? Couldn't it be causing both health problems and low cholesterol?

**A:** Yes. The alcohol/low-cholesterol connection is this: In some drinkers, alcohol damages the liver, hampering its cholesterol-making capacity and resulting in lower blood-cholesterol levels. In fact, doctors who do autopsies say that people who die of cirrhosis of the liver often have remarkably clean arteries.

But the association seems to be weak, since cholesterol drops only when someone has severe liver disease, and the association remains even when people with severe liver disease are excluded.

**Q:** What about the side effects of cholesterol-lowering drugs? Could they be causing the increase in deaths?

**A:** In some cases, yes. Both gemfibrozil and clofibrate, drugs we talk more about in Chapter 6, have been implicated in some deaths, especially those involving liver and gallbladder disease.

**Q:** Any other explanations?

**A:** Some researchers speculate that, since cholesterol is a vital part of cell membranes and hormones in the body, too-low levels could cause a wide variety of diseases and dysfunction, Jacobs says.

**Q:** Such as?

**A:** Everything from cancer to stroke to mood or behavior problems that might lead to accidents or suicide.

**Q:** Any proof that that's actually the case?

**A:** Only a bit, and mostly in monkeys on cholesterol-lowering diets.

For instance, monkeys fed a moderately low-fat diet, getting 30 percent of calories from fat, have reduced levels of **serotonin**, a brain neurotransmitter that helps inhibit impulsive behavior, explains Jay R. Kaplan, Ph.D., of the Bowman-Gray School of Medicine, Wake Forest University, Winston-Salem, North Carolina. These monkeys are also more aggressive than monkeys fed a high-fat diet, even though they get plenty to eat.

**Q:** But do men who lower their cholesterol a lot get impulsive or aggressive?

**A:** "No one knows. No one has ever measured it," says Matthew F. Muldoon, M.D., M.P.H., a researcher at the University of Pittsburgh's Center for Clinical Pharmacology who has just started a study on the effects of cholesterol lowering on mood. The study will include measurements of mood-altering brain chemicals such as serotonin in men and women taking cholesterol-lowering drugs. It will also examine brain biochemistry in people with naturally low cholesterol levels.

"We're hoping this study will provide some answers to this question," Muldoon says.

**Q:** So there's no proof that cholesterol-lowering drugs or a low-fat diet causes mood problems in humans?

**A:** While there's some speculation that could happen, there's no proof—not yet anyway. And there's some evidence to the contrary.

"We just completed a cholesterol-lowering dietary-intervention study in children," says Kwiterovich of Johns Hopkins. "Half the children were on a special cholesterol-lowering diet; the other half got usual care. We assessed psychosocial development, depression and a whole bunch of things, and there was no difference between the two groups.

"We regularly get people's total cholesterol down to 160 mg/dL and their LDL down to about 100 mg/dL, and we find no evidence of depression," adds Stephen Inkeles, M.D., a staff physician at the Pritikin Longevity Center, Santa Monica, California, and a clinical professor of medicine at UCLA. "Even at this low level, there seems to be plenty of cholesterol to go around to make all the hormones a person needs."

**Q:** **So what does all this mean? Should I just not worry about my cholesterol and go have a cheeseburger?**

**A:** No, it doesn't mean that at all. But here's where it pays off to weigh risks versus benefits.

"Studies clearly show that in people with the highest risk—those who already have signs of coronary heart disease—the benefits of cholesterol lowering outweigh any possible risks," Jacobs says. "It would be irresponsible to tell people in this category not to do everything they need to—diet, and drugs, if necessary—to lower their cholesterol into a low-risk range." All the doctors we talked with agree this is sound advice.

**Q:** **But what if I simply have high or borderline-high cholesterol but no signs of heart disease? What should I do then?**

**A:** Here again, you have to carefully consider your potential benefits and risks and get your doctor's advice, experts say. Find out what risks for heart disease you can eliminate with no risk and little cost—quit smoking, lose weight and start exercising.

Changes in diet are now considered the preferred choice for lowering cholesterol for most people, says Inkeles. "Dietary changes have never had anything but beneficial side effects—like preventing cancer, diabetes, obesity and high blood pressure."

If you have a serious inherited cholesterol disorder, however, you may benefit from taking drugs to reduce your cholesterol, even though you have no signs of heart disease.

**Q:** **Does this mean there is an ideal cholesterol range—that you shouldn't try to go below a certain level?**

**A:** As we said earlier, most problems seem to occur at cholesterol levels 160 mg/dL or lower. That low level is achievable on either diet or drug therapy, or both. So you might want to aim for not much lower than that, with LDL levels of about 100 mg/dL, Inkeles says.

# 5 CHOLESTEROL- AND TRIGLYCERIDE- LOWERING DIETS

## REVERSING CORONARY-ARTERY DISEASE

**Q:** I know we'll be discussing the types of diet that help prevent heart disease, but I'd like to know first—can you actually *reverse* atherosclerosis if your blood levels of cholesterol are low enough?

**A:** Yes. A number of studies have found it's possible for some people to undo at least some of the artery-choking effects of years of burgers and French fries.

**Q:** What exactly are these studies showing?

**A:** One report, by researchers at the Lawrence Berkeley Laboratory at the University of California, looked at 10 of these studies, involving a total of 2,095 men and women. All 10 were randomized, controlled clinical trials using coronary **arteriography** (a test that can measure obstruction inside an artery) to assess the effects of cholesterol lowering. In other words, these studies were carefully designed and had an objective means of measuring results.

**Q:** So what did they find?

**A:** The researchers found that artery obstruction diminished in nearly one-third of those on an aggressive cholesterol-lowering treatment program. Only one-tenth of people who received modest interventions (a moderately reduced-fat diet and, usually, bile acid binding resins, which we talk about in Chapter 6) had diminished obstructions.

And in the six studies that reported cardiovascular events, there were fewer heart attacks among the aggressively treated individuals.

**Q:** What's that mean—aggressive treatment? Were these studies lowering cholesterol a lot?

**A:** Yes. They reduced it further than is done in most studies, achieving, on average, a 28 percent reduction in LDL cholesterol, an 11 percent reduction in triglycerides and an 11 percent increase in HDL cholesterol, compared with control groups, which lowered total blood cholesterol an average of 10 percent.

**Q:** What's that look like in terms of actual cholesterol levels?

**A:** In these studies, the numbers averaged out this way for the aggressively treated group: Triglycerides averaged 172 mg/dL to start. They dropped 11 percent, to 153 mg/dL; total cholesterol was 276 mg/dL. It dropped 26 percent to 205 mg/dL. LDL, at 196 mg/dL, dropped 36 percent to 125 mg/dL; and HDL, initially at 42 mg/dL, rose 17 percent to 49 mg/dL.

**Q:** But those figures aren't as low as the target goals you cited earlier for people with heart disease. Why not?

**A:** That's true, they're not. Some doctors believe a large percentage drop in cholesterol helps reverse the course of heart disease as much as reaching a particular target figure. And if these people could go even lower, perhaps they'd see even more reversal, says Stephen Inkeles, M.D., of the Pritikin Longevity Center.

"We look most at LDL cholesterol, and we like to see it below 100 mg/dL," he says. That's usually along with total cholesterol of 160 to 180 mg/dL and HDL levels above 30 mg/dL.

**Q:** Did these studies use drugs or diet?

**A:** Four of the studies used diet, exercise and the like; seven used single- and multiple-drug therapy combined with diet.

**Q:** Which worked better—drugs or diet?

**A:** Two of the nondrug studies worked better than any of the drug studies. Patients in both had about a 40 percent greater rate of regression than those in a control group—they were 40 percent more likely to have shrinkage of artery blockages.

**Q:** What kinds of diets are being used in these studies?

**A:** The studies by Dean Ornish, M.D., director of the Preventive Medicine Research Center, Sausalito, California, showed an impressive reversal of blockages. The diet used there is mostly vegetarian, rich in vegetables, fruits, grains and beans, which are naturally low in fat and high in fiber. Total fat intake is cut to 10 percent of calories. Foods of animal origin are limited to nonfat dairy products and egg whites and one serving of lean poultry or fish (no red meat) three times a week.

This diet has about 20 grams of fat for an 1,800-calorie diet. On this diet, you avoid every known booster of blood cholesterol—too much total fat, saturated fat, **trans fatty acids** and dietary cholesterol. (We discuss these dietary components later in this chapter.) You also get lots of soluble fiber, which also helps reduce cholesterol. And you get lots of vitamins E and C and **beta-carotene**, a vitamin A precursor. All three of these nutrients, especially vitamin E, appear to help stop an early stage of atherosclerosis we talked about in Chapter 1: the oxidation of LDL cholesterol.

**Q:** How long does it take to see results using such a diet?

**A:** Less time than you might think. Most studies saw plaque shrinkage in only two to four years —that's considered a relatively short period of time as far as heart disease goes.

**Q:** How exactly do these fatty plaques shrink?

**A:** The details aren't entirely known, but only "active" plaques—those that still contain macrophages and aren't completely covered with fiber and minerals—seem to be able to reverse.

"And once cholesterol levels get low enough in the blood, LDL cholesterol may move out of the plaque into an area of lower concentration, where, apparently, it is picked up by HDL and carried to the liver," Inkeles says.

**Q:** How much shrinkage can you expect to see in these fatty plaques?

**A:** Studies show that, most of the time, they shrink just a bit—in most cases, less than 2 percent.

**Q:** Two percent? Then what's the big deal?

**A:** For starters, it means the plaque *isn't* increasing in size by about 1.5 percent, as it otherwise would do during this period of time. And even with

this small amount of shrinkage, there's still an improvement in symptoms—more than one might expect.
That's apparently because the reduction in cholesterol
levels also improves the function of endothelial cells—
the cells lining arteries.

**Q: What's that got to do with it?**

**A:** These cells do more than act as a tube for blood,
Inkeles explains. "They secrete biochemicals
that play an important role in the health of the circu-
latory system."

For one thing, they secrete chemicals that help blood
vessels relax. That can help protect against the **vasospasms** that can cause angina, or chest pains caused by
a lack of oxygen to the heart.

Endothelial cells also secrete chemicals that inhibit
platelet stickiness and so reduce the tendency for
clotting at the site of a plaque. "These functions are
very important when it comes to preventing heart
attacks," according to Inkeles.

**Q: Did any of these studies include people
who'd had bypass surgery?**

**A:** Yes. The Cholesterol Lowering and Atherosclerosis Study studied men who'd had bypass
surgery. Half the men took two cholesterol-lowering
drugs we talk about in Chapter 6, colestipol and nicotinic acid. The other half took a placebo. The men
were followed for a total of four years.

**Q:** So what happened? Did the drugs help keep the grafts open?

**A:** In a word, yes. And it was most apparent the drugs were helpful in the fourth year of the study. Arteriograms done then showed nonprogression of fatty plaques in 52 percent of men taking the two drugs, compared with only 15 percent of men in the placebo group. And regression was 18 percent in the treated group, compared with 6 percent in those taking the placebo.

People in the treatment group were also less likely to have new fatty plaques in nongrafted, or "native," arteries. Fourteen percent had new plaques, compared with 40 percent in the control group. And in their bypass grafts, new plaques occurred in only 14 percent, versus 38 percent in the placebo group.

**Q:** That's important, isn't it? I've heard that bypasses can clog up fast.

**A:** They can. The rate of progression of fatty plaques is reported to be three to six times greater in grafted versus nongrafted vessels. Without intervention to lower cholesterol, 45 percent of grafts are seriously clogged up five years after bypass surgery.

**Q:** How much of a role does diet really play in high cholesterol? Or are bad genes to blame?

**A:** It's true that some people have inherited disorders that keep their cholesterol or triglyceride levels high despite dietary measures. And some people

can eat just about anything they want and never have a
problem with high cholesterol. But most people fall
somewhere in between.

"Some people will drop their cholesterol 100 mg/dL
or more when they go on a low-fat, low-cholesterol
diet, and some just won't budge at all," says Wayne
Callaway, M.D., a specialist in metabolism and nutrition
at George Washington University, Washington, D.C.
"No one really knows exactly how many people are
truly resistant to dietary changes, since most studies
simply look at average declines in cholesterol and the
people who don't respond are often simply assumed to
be noncompliant—not following the diet."

One type of study—metabolic ward studies, where
people are, basically, paid captives whose every bite is
controlled—shows that it is indeed the rare individual
who doesn't respond at all to diet. The only way to
know how much you'll respond to a diet—and if you
are truly diet-resistant—is to try a particular diet and
see if it lowers your cholesterol, Callaway says. If you
already have heart disease, a low-fat diet may also reduce your symptoms of chest pain, fatigue and so forth.

## Q: So most people can reduce their cholesterol with diet?

## A: Yes. Most doctors today believe too much fat in the diet, especially saturated fat, plays a major
role in jacking up cholesterol levels.

They point to the fact that in countries where a low-
saturated-fat diet is the norm, such as the traditional
rice-and-fish Japanese menu, cholesterol levels *average*
160 mg/dL or lower. But when these people switch to a
high-fat American diet, their cholesterol levels rise,
along with their risk for heart disease.

Doctors also cite the fact that most people, if they are willing to work at it, can lower their cholesterol on a sufficiently low-fat diet and, in some cases, can even slightly shrink artery blockages.

**Q:** So is that what causes high blood cholesterol? A high-fat diet?

**A:** It's certainly one big reason. We all need some fat in our diets. But we need only enough to supply our bodies with adequate amounts of two essential fatty acids, **linoleic** and **linolenic acid**. It's possible to get that amount by eating 15 to 25 grams of the kind of fat that characterizes the American diet. However, the best sources are vegetable oils—**canola**, safflower, corn, soybean and cottonseed.

But most people, in the United States anyway, get 80 to 100 grams of fat, or about one-third of a cup, a day. That's the problem—too much fat, especially saturated fats.

Some doctors think the high-saturated-fat American diet puts everyone here at risk for coronary-artery disease, and to some extent, statistics back them up. If you don't die of cancer, chances are pretty good that you'll die of a heart attack, stroke or some other problem associated with atherosclerosis.

**Q:** But don't other things besides a high-fat diet elevate cholesterol?

**A:** Yes. Stress can raise cholesterol levels, as can liver, gallbladder or thyroid disorders. And as we said earlier, high cholesterol is only one of numerous

risk factors for heart disease. Since most people in the
United States have at least borderline-high cholesterol—
200 to 249 mg/dL—those other risk factors play an
important role in determining who's most at risk.

But since high cholesterol seems so closely linked
with a high-fat diet, which can easily enough be pared
down, and since a high-fat diet is associated with
another big killer, cancer, a lot of the emphasis these
days is on cutting out fat, especially saturated fats and
**hydrogenated fats**, which we'll explain shortly.

# Q: Exactly what kind of diet is recommended to reduce cholesterol?

# A: These days, there are a number of diets; some
cut fat only a bit, to just under 30 percent of
calories, and are recommended for anyone wanting to
eat healthfully, high cholesterol or not. Others slash fat
to the bone, to 10 percent or less of calories, and are
geared toward people who already have blocked arteries
or very high cholesterol. In between are moderately
reduced-fat diets, with 15 to 25 percent of calories
from fat.

And some doctors recommend a Mediterranean-style
diet, which has 25 to 30 percent total fat, mostly from
a **monounsaturated fat**, olive oil, but which is low in
saturated fats.

## SATURATED FATS

**Q:** Before we go any further, explain these fats to me. What exactly are saturated fats?

**A:** Saturated fats are hard at room temperature. Saturated fats are the main component of the white marbling in meats, of the visible fat around meat, of butter and cheese and of whole milk and ice cream. Saturated fats are also plentiful in coconut oil, palm-kernel oil and palm oil, three plant oils widely used in commercial baked goods like cookies and crackers.

**Q:** Why are saturated fats supposed to be so bad for you?

**A:** Most saturated fats—although there are exceptions—raise blood levels of LDL cholesterol. They do this, apparently, by increasing cholesterol production in the body and by inhibiting the uptake of LDL cholesterol by cells. Cells can't take in cholesterol circulating in the blood as easily when you're eating a high-saturated fat diet as they can when saturated-fat intake is low.

**Q:** Do most people eat a lot of saturated fat?

**A:** You bet. The average American can pack it away. Most of us get about 14 percent of our calories from saturated fat. Researchers have calculated that each percent of calories from saturated fat raises total cholesterol by approximately 2.3 mg/dL. In fact, saturated fats raise blood cholesterol levels more than dietary cholesterol does.

**Q:** But I've heard saturated fat is not so bad. What's the story?

**A:** You may have heard about a component of saturated fat, called **stearic acid**, found in red meat and chocolate. Recent research suggests that stearic acid does not elevate LDL cholesterol. However, stearic acid is only one small fraction of the fatty acids found in saturated fats.

**Q:** What is the main fatty acid in saturated fats?

**A:** It's a component called **palmitic acid**, and it's the predominant saturated fat in red meat, chicken and butter fat, and even in certain vegetable fats and oils, such as palm and cocoa butter.

**Q:** Does palmitic acid raise LDL cholesterol levels?

**A:** Yes, definitely. And palmitic acid is the component of saturated fat people in the United States eat most. Most people get 8 to 10 percent of their calories from palmitic acid, an amount calculated to raise their cholesterol 22 to 27 mg/dL.

## TRANS FATTY ACIDS

**Q:** **What about fats in margarine and vegetable shortening (Crisco)? They're hard at room temperature.**

**A:** These products contain hydrogenated vegetable oil. Manufacturers have added hydrogen, which makes these products harder and also helps to keep them from becoming rancid. Hydrogenated vegetable oils contain trans fatty acids.

**Q:** **Do trans fatty acids raise cholesterol levels?**

**A:** Yes. They act in ways similar to saturated fats. About half a dozen studies have found that the amount of trans fatty acids found in the typical American diet—8 to 15 grams a day—can raise LDL cholesterol levels as much as saturated fats do. One study also found that about twice that amount— 30 grams a day—lowered HDL cholesterol by about 13 percent, which translates into a 26 percent increase in risk for heart disease.

A study by Harvard University researchers, for instance, found women with the highest intake of trans fatty acids (about 6 grams a day) were 50 percent more likely to develop heart disease than those who took in less than half as much. They also found that those who ate four or more teaspoons a day of margarine raised their risks of heart problems by 66 percent, compared with those who ate less than one teaspoon a month. The researchers saw no such risk from butter.

Another researcher, Bruce Holub, a nutritional biochemist at the University of Guelph, in Ontario, compared how trans fatty acids stacked up against saturated fats, such as those found in butter. Considering LDL alone, he estimated that 1 gram of trans fat has 0.7 times the cholesterol-raising capacity of the same amount of saturated fat. Factoring a dip in HDL levels would boost that number to 1.3. However, many researchers believe most people don't eat enough trans fatty acids to affect HDL levels.

**Q:** **So what should I do? Not eat margarine? Eat butter instead?**

**A:** Stick with soft-spread margarines, which have fewer trans fatty acids per gram than stick margarines. Look for those that list water or a liquid vegetable oil as the first ingredient. If you've been melting margarine to use in cooking, switch to olive or canola oil instead.

If you prefer butter to margarine and your risk for heart disease is low, you can use butter sparingly without guilt.

**Q:** **How much trans fatty acid can I safely eat?**

**A:** Researchers are saying now you should lump trans fatty acids with saturated fats. That means you shouldn't be getting more than 10 percent of your calories from saturated fats and trans fatty acids combined—from about 18 to 26 grams a day for most people.

Margarine, fried fast foods and packaged baked goods are the main sources of trans fatty acids for most people in the United States. A fried fish sandwich has 8.2 grams; a fried fruit pie, 5.9 grams; a plain doughnut, 3.2 grams; a small order of French fries, 3.6 grams; a tablespoon of stick margarine, 3.1; and two small cookies, 1.7.

Baked goods marketed as "made with all vegetable oils" may be high in trans fatty acids. If so, they'll list hydrogenated or partially hydrogenated vegetable oil as one of the first three ingredients on their labels.

## POLYUNSATURATED FATS

**Q:** What are **polyunsaturated fats?**

**A:** Polyunsaturated fats are liquid at room temperature and are the predominant fat in common vegetable oils, such as corn, safflower, sunflower, cottonseed, soybean and walnut oils.

One reason for the corn-oil craze back in the 1970s is that some studies suggested that unsaturated fats such as the "polys" can actually reduce total cholesterol. In fact, however, research now suggests that while polyunsaturated fats do not raise harmful LDL cholesterol, they do lower helpful HDL cholesterol, hardly a desirable result. And in animals, at least, diets high in polyunsaturated fats may increase the risk of cancer.

These fats are also easily oxidized. That means that, in your body, incorporated into LDL, they react with oxygen in a chemical reaction that's identical to butter becoming rancid. As we explained in Chapter 1, oxidation of LDL cholesterol is thought to be one of the initial stages of fatty plaque development in arteries. So it's good to keep your intake of these fats fairly low. We talk about the **antioxidant** properties of some vitamins, especially vitamin E, later in this chapter.

## MONOUNSATURATED FATS

**Q:** What about the monounsaturated fats you mentioned earlier?

**A:** Monounsaturated fats are also liquid at room temperature. Olive oil is 77 percent mono-unsaturated fat. Canola (rapeseed) oil is 58 percent monounsaturated. Peanut oil contains some mono-unsaturated fat.

**Q:** I've heard that olive oil is supposed to be good to use if you are concerned about high cholesterol. Is it?

**A:** It does seem to be. In studies, switching from saturated to monounsaturated fats, such as olive oil, made LDL cholesterol drop without also dropping HDL cholesterol. Studies from French researchers suggest that an olive-oil-rich diet changes the structure of HDL, increasing its capacity to transport LDL cholesterol to the liver for disposal.

Other studies indicate that olive oil is less prone to oxidation than polyunsaturated fat. In "olive-oil belt" countries—Greece and Italy, for instance—people eat a diet that's fairly high in monounsaturated fats but low in saturated fats. And their risk for heart disease is a fraction of that in the United States. Olive oil contains more heart-protective vitamin E than other vegetable oils, which may be one reason for its apparent heart-healthy benefits.

## DIETARY CHOLESTEROL

**Q:** You haven't mentioned anything about dietary cholesterol. Isn't it important to limit the amount of cholesterol you eat, too?

**A:** That used to be the party line, but it's now open to debate. The American Heart Association still says that too much dietary cholesterol does raise your blood level of cholesterol, and recommends you get no more than 300 milligrams of cholesterol a day from foods. That comes to no more than three to four large eggs a week, and for most people it means cutting back. On an average day, a man consumes about 450 milligrams and a woman about 320 milligrams of cholesterol.

**Q:** Why is this open to debate? Doesn't eating too much cholesterol raise your cholesterol?

**A:** Not for everybody and, apparently, not for a lot of people, once total fat and saturated fat are pared away, says Callaway.

"Studies indicate that roughly one-third of people have a significant drop in blood cholesterol if they lower dietary cholesterol," he says. Those people will benefit by reducing their dietary intake of cholesterol by 200 milligrams or more.

There's also a small group that can eat practically unlimited amounts of cholesterol a day with no change in blood cholesterol. And then there's a large middle group of people who have minor responses to dietary cholesterol.


**Q:** **How will I know if I should cut back?**


**A:** "If your total, HDL and LDL cholesterol are all within safe ranges, then I'd say keep eating however you've been eating," Callaway says. However, if your cholesterol is high, you'll want to keep your egg intake to three to four a week, since eggs contain a fair amount of saturated fat in addition to cholesterol.


**Q:** **How much cholesterol actually is in an egg?**


**A:** An egg yolk from one large egg contains a little more than 200 milligrams of cholesterol and about 5.1 grams of fat; 1.6 grams of it is saturated fat. Many baked goods contain egg yolks, unless the label says they are cholesterol-free.

**Q:** I've heard liver contains a lot of cholesterol. Is it okay to eat?

**A:** If your cholesterol is normal, there's no reason not to have a serving of liver now and then, since liver is packed with nutrients, such as vitamins B12 and A. It's true that liver and other organ meats, such as sweetbreads (thymus or pancreas), brain and heart, are loaded with cholesterol: Three ounces of beef liver yield 331 milligrams of cholesterol. That same serving contains only 1.6 grams of saturated fat, but if your cholesterol is high, you're better off avoiding it.

## FISH AND SHELLFISH

**Q:** What about shrimp? I've heard they're not so bad to eat.

**A:** Certain shellfish, such as shrimp and crayfish, are higher in cholesterol than most other types of fish and seafood. But they're lower in total fat and saturated fat than most meats and poultry. Three ounces of shrimp have 166 milligrams of cholesterol, but only 0.25 milligrams of saturated fat. So you're better off eating shrimp than those other foods. Just make sure you eat them steamed, not fried, which adds lots of fat.

**Q:** What about other seafoods?

**A:** Surf beats turf, hands down. In fact, a study by researchers at the University of Washington found that men who substituted oysters, crab, lobster, clams, mussels or scallops for cheese, meat or eggs had a reduction in LDL cholesterol and an increase in HDL cholesterol. Shrimp and squid didn't improve cholesterol levels, but they didn't harm them, either.

**Q:** Isn't there some sort of fat in fish that's supposed to be good for your heart?

**A:** Yes. Some fatty fish such as salmon, mackerel, tuna and some shellfish contain **omega-3 fatty acids**. Smaller amounts are also found in whole grains, beans, seaweed and soybean products. And purslane, a fleshy, low-growing weed sometimes used in salads, contains high amounts of omega-3 fatty acid.

These fats are polyunsaturated but their chemical structure is different from other polyunsaturated fats.

**Q:** What do omega-3 fatty acids do for cholesterol levels?

**A:** They don't seem to have any effect at all on cholesterol, but they do lower triglycerides in some people. That apparently, however, is not their main benefit when it comes to preventing heart disease. Fish oils' biggest benefit may come from their ability to reduce a process called **platelet aggregation**, an important part of blood clotting. A diet rich in fish oils helps keep platelets from sticking together and from sticking to artery walls, just as does aspirin. This means blood clots are less likely to form on top of cholesterol deposits.

Fish oils also dampen the body's inflammatory response, which may play a role in the early development of atherosclerosis.

**Q:** **Does this mean I should eat more fish?**

**A:** Certainly, substitute fish for some of the beef or chicken you are eating. Some experts recommend up to three fish meals a week. However, since even fish contains some saturated fat, if you're going super-low-fat, you'll want to stick mostly with the vegetable sources of omega-3 fatty acids—seaweeds, soybeans and purslane.

**Q:** **What about fish-oil capsules?**

**A:** You're better off eating fish. Three ounces of mackerel, tuna or salmon offer 2,500 milligrams of omega-3 fatty acids, while a typical omega-3 capsule contains only 300 milligrams. These capsules are made up mostly of other fats.

## SUGAR AND ALCOHOL

**Q:** You've talked about foods that affect cholesterol levels. Are there any foods that affect triglyceride levels?

**A:** Alcohol and simple carbohydrates, such as sugar, raise triglyceride levels directly through their effect on the liver. They cause the liver to produce VLDL cholesterol, which then breaks down to form triglycerides and LDL cholesterol.

**Q:** Alcohol raises triglyceride levels? But I've heard wine is supposed to be good for your heart. Is it?

**A:** Studies show that moderate drinking—one drink a day for women, two for men—is associated with a reduced risk of heart attacks. In fact, the reduction in risk is comparable to the protection conferred by taking low-dose aspirin, stopping smoking, exercising or losing weight.

**Q:** How does alcohol work?

**A:** One or two drinks a day increase HDL cholesterol 5 to 10 mg/dL. Alcohol also reduces the tendency for blood clotting, a factor in clogged arteries. And the substances that give red wine and grape juice their deep rich color may also inhibit the oxidation of LDL cholesterol in artery walls.

**Q:** But should I avoid alcohol if my triglyceride levels are high?

**A:** Yes. In that case, the potential negative effects—higher-than-ever triglyceride levels— outweigh any potential benefits.

**Q:** Don't fats also raise triglyceride levels?

**A:** Dietary fats can also raise triglyceride levels right after a meal, but this triglyceride is a special, temporary form. It is transported in **chylomicrons**, tiny spheres of fat that go directly from the intestines to the liver and that are thought to have no direct effect on the development of atherosclerosis.

One reason triglyceride levels sometimes rise when people go on low-fat diets is that they're substituting carbohydrates and sugar for fat. Some people with high triglyceride levels, and especially those with diabetes, actually do better in terms of cholesterol and blood-sugar control if they eat a diet that contains a bit more monounsaturated fat and fewer carbohydrates and sugar.

## WHEN TO GO LOW-FAT

**Q:** How high does someone's cholesterol need to be for a doctor to recommend a low-fat diet?

**A:** The National Cholesterol Education Program has set these guidelines for starting a low-fat diet:

• If you don't have heart disease and have fewer than two of the risk factors we listed in Table B, Chapter 1, such as smoking or high blood pressure, it's recommended you start dietary therapy if your LDL cholesterol is more than 160 mg/dL. Your goal should be to keep your LDL cholesterol below 160 mg/dL and your total cholesterol lower than 240 mg/dL.

• If you don't have heart disease but have two or more risk factors, it's recommended you start dietary therapy if your LDL cholesterol is more than 130 mg/dL. Your goal is to keep your LDL cholesterol under 130 mg/dL and your total cholesterol under 200 mg/dL.

• If you have heart disease, dietary intervention is supposed to start when LDL cholesterol is higher than 100 mg/dL; the goal is to keep LDL under 100 mg/dL and total cholesterol at 160 mg/dL or less.

**Q:** What kinds of dietary recommendations are made?

**A:** Doctors are advised to start their patients on one of two diets offered by the National Cholesterol Education Program. They're similar to the American Heart Association's Step I and Step II diets.

The Step I Diet calls for 30 percent or less of calories from fat, with 8 to 10 percent of calories from saturated fat and 300 milligrams or less of cholesterol a day. Your doctor may suggest this diet to "ease you into" a lower-fat diet, especially if your current diet is very high in fat.

**Q:** How's the Step I diet different from how most people eat?

**A:** Studies show that the typical American diet has 40 percent of calories from fat, with 13 to 14 percent saturated, 14 percent monounsaturated and 8 percent polyunsaturated fats, and a cholesterol intake of 350 to 400 milligrams a day.

**Q:** So what's the Step I diet look like on a plate?

**A:** The Step I diet is not particularly stringent. It's the kind of diet most doctors these days encourage everyone to follow. For 1,800 calories a day, the amount most women eat, a Step I diet looks like this:

**Breakfast:** ½ medium bagel, 1 teaspoon low-fat cream cheese, 1 cup shredded wheat cereal, 1 small banana, 1 cup 1 percent milk, ¾ cup orange juice and 1 cup coffee with 2 tablespoons 1 percent milk.

**Lunch:** ½ cup minestrone soup, 1 roast beef sandwich
(made with 2 slices whole-wheat bread, 3 ounces
lean roast beef, ¾ ounce low-fat American
cheese, 1 leaf lettuce, 3 slices tomato and
2 teaspoons low-fat mayonnaise), 1 medium
apple and 1 cup water.

**Dinner:** 3 ounces salmon, 1 medium baked potato
with 1 teaspoon tub margarine, ½ cup green
beans with ½ teaspoon tub margarine, ½ cup
carrots with ½ teaspoon tub margarine, one
medium white dinner roll with 1 teaspoon
tub margarine, ½ cup ice milk and 1 cup
unsweetened iced tea.

**Snack:** 2 cups popcorn with 1 teaspoon tub margarine.

**Q:** **What kind of results can you expect to see
on this sort of diet?**

**A:** A wide range—from 3 to 14 percent cholesterol
reduction—depending on your cholesterol level
and your eating habits when you start the diet. The
higher your blood cholesterol level and the fattier your
regular diet, the more of a drop you're likely see on a
low-fat diet. Men eating an average American diet can
expect a drop of 5 to 7 percent in total cholesterol.
Women may see slightly less of a drop.

The only way to find out for sure how you're going
to respond is to try the diet. Incidentally, people who
drop excess weight on a low-fat diet often see much
better cholesterol-lowering results. We talk more about
obesity and heart disease in Chapter 7.

**Q:** How long do I need to be on this diet before I know if it's going to work?

**A:** A minimum of three months. Then you'll have your blood level of cholesterol checked to see how much it's dropped.

**Q:** What happens if this diet doesn't do the trick for me?

**A:** The Step I diet has been criticized for not cutting fat enough to do a whole lot of good. So it may not be restrictive enough for you.

If you're actually sticking with the diet—and any exercise and weight-loss recommendations your doctor has advised—and your cholesterol still hasn't dropped enough after three months, your doctor will most likely recommend a diet that is more restrictive in saturated fat—the Step II diet.

**Q:** How is it different from the Step I diet?

**A:** This diet reduces saturated fat to less than 7 percent of calories and cholesterol to less than 200 milligrams a day. But it allows more servings of vegetable fats and oils (poly- and monounsaturated fats), thereby keeping total calories from fat the same as for Step I—30 percent. Most of the reduction in saturated fat comes from smaller portions of meat and cheese and skim milk instead of 1 percent.

# Q: So what's that look like on a plate?

# A: For 1,800 calories a day, a Step II daily menu plan looks like this:

**Breakfast:** ½ medium bagel, 1 teaspoon jelly, 1 cup shredded wheat cereal, 1 small banana, 1 cup skim milk, 1 cup orange juice and 1 cup coffee with 2 tablespoons skim milk.

**Lunch:** ½ cup minestrone soup, 1 roast beef sandwich (made with 2 slices whole-wheat bread, 2 ounces lean roast beef, ¾ ounces low-fat American cheese, 1 leaf lettuce, 3 slices tomato and 2 teaspoons tub margarine), 1 medium apple and 1 cup water.

**Dinner:** 3 ounces flounder, 1 teaspoon vegetable oil, 1 medium baked potato with 1 teaspoon tub margarine, ½ cup green beans with ½ teaspoon tub margarine, ½ cup carrots with ½ teaspoon tub margarine, one medium white dinner roll with 1 teaspoon tub margarine, ½ cup low-fat frozen yogurt and 1 cup unsweetened iced tea.

**Snack:** 3 cups popcorn with 2 teaspoons tub margarine.

# Q: What kind of results can you expect to see on the Step II diet?

# A: Another 3 to 7 percent reduction in total cholesterol, which, added to the 3 to 14 percent achieved by the Step I diet, results in a total of 6 to 20 percent.

**Q:** How long a time are people supposed to give this diet?

**A:** The National Cholesterol Education Guidelines call for a minimum of three months, but some doctors say it takes people longer than that to really learn how to follow such a diet. They recommend giving it a year, if possible, before considering the use of drugs.

**Q:** What's the best way to figure out how to make such a diet work for me?

**A:** Get expert help, right from the start, Callaway says. See a dietitian or nutritionist who can analyze your current diet and figure out exactly what it will take to get you into compliance with a lower-fat diet. Ask your doctor for a referral to a dietitian or nutritionist, or simply seek one out on your own.

**Q:** Can't my doctor just do this herself?

**A:** Most doctors don't take the time and are not nearly as well educated as dietitians when it comes to the nitty-gritty details of fat cutting.
"I think depending on your primary-care physician for dietary advice is a waste of time, because her knowledge is very superficial," Callaway says. "A person comes back to the doctor in three months, his cholesterol level shows the diet hasn't worked and the doctor puts him on drugs. The fact is, in that situation, you just don't know for sure if diet would have helped

because they haven't kept food records, they haven't reviewed them with a dietitian, and no one, not even the patient, is sure whether or not the diet was actually followed."

**Q:** What if my doctor doesn't work with a dietitian?

**A:** Change doctors. Or call a local hospital and ask to speak to a dietitian on staff. You can also write to the American Dietetic Association, 430 Michigan Ave., Chicago, IL 60611 to get the names of dietitians in your area who specialize in cholesterol-lowering diets. Send your request, along with a self-addressed stamped envelope and $1.

## VITAMINS AND HEART DISEASE

**Q:** Can't certain vitamins help prevent heart disease?

**A:** Apparently so. As we already mentioned, vitamins E and C and beta-carotene all appear to help reduce risk of heart disease. All three function as antioxidants. They help to protect LDL cholesterol from oxidation—reacting with oxygen—a first step in the process of atherosclerosis, which we discussed in Chapter 1. Vitamin E does the job directly, by entering the artery wall in the same lipoprotein particles that carry LDL cholesterol. The role of beta-carotene is less understood; it may also enter the arterial wall to provide protection against oxidation. And vitamin C's main role in all this is apparently to protect vitamin E so it can keep on protecting against the LDL cholesterol.

**Q:** But do these vitamins actually lower cholesterol?

**A:** So far only one, vitamin C, seems to influence cholesterol levels. Researchers at Tufts University Center for Aging, in Boston, found that large doses of vitamin C lower LDL and raise HDL levels. They found that women's HDL peaked when vitamin C in their blood reached 1 mg/dL. For men, HDL continued to rise parallel with vitamin C, while the LDL cholesterol dropped significantly.

For women, the amount needed to reach that blood concentration was about 90 milligrams—the amount in six ounces of orange juice. Men reached the same concentration when they had consumed about 150 milligrams. Since vitamin C levels drop quickly after a large dose, most experts suggest you take at least two doses a day.

It's possible to get enough vitamin C and beta-carotene in your diet by eating lots of fruits and vegetables, especially citrus fruits and dark leafy greens. To get protective amounts of vitamin E—at least 400 international units a day, according to one study—requires supplements.

# Q: Do any other vitamins or minerals affect cholesterol or triglycerides?

A: We talk about the B vitamin niacin, which is often used in large doses to lower cholesterol, in our section on drugs. The form of niacin that's actually used is nicotinic acid.

One other nutrient deserves mention here—chromium. Chromium deficiency seems to play a role in the development of atherosclerosis. Several studies have shown that one form of this trace mineral, chromium picolinate, may help reduce high LDL cholesterol and triglyceride levels and raise HDL levels. And it seems especially helpful to people with diabetes.

In one study, for instance, moderately obese noninsulin-dependent diabetics who took 200 micrograms of chromium picolinate every day for two months experienced a 17.4 percent drop in triglyceride levels. There was no change while they were taking a blank, look-alike drug.

# Q: Can I get that much chromium from foods?

A: It's not likely. Most Americans don't get enough chromium in their diets. Brewer's yeast, wheat germ, broccoli, cheese and prunes are good sources, but you may need to take supplements to reach cholesterol-lowering amounts. Some doctors recommend up to 200 mcg. three times a day to their patients with diabetes and high triglycerides. However, these high doses deserve medical supervision. Although no toxicity has been reported in amounts up to 600 mcg. a day, chromium can be toxic.

**Q:** **What about the B vitamin folic acid? I've read that it's supposed to help prevent heart disease. Does it?**

**A:** Apparently so. Low levels of folic acid increase blood levels of homocysteine, an amino acid that may damage the endothelial cells lining artery walls, making it easier for LDL cholesterol to deposit there. Folic acid does not affect blood cholesterol levels, however.

To keep homocysteine levels low, researchers suggest you get 400 mcg. a day of folic acid. Most people get about half that amount in a normal daily diet. To reach 400 mcg., include orange juice (one cup has 110 mcg.), leafy greens (one cup of raw spinach has 130 mcg.) and vitamin supplements or folate-fortified breakfast cereals in your daily diet.

# 6 CHOLESTEROL- AND TRIGLYCERIDE- LOWERING DRUGS

## GUIDELINES FOR USING CHOLESTEROL-LOWERING DRUGS

**Q:** When are drugs used to lower cholesterol or triglyceride levels?

**A:** Drugs are meant to be used only *in addition to* lifestyle measures, such as eating a low-fat diet, stopping smoking, losing weight and exercising. That's true even in people who need drugs from the start because of high initial cholesterol or triglyceride levels. It's also true for people with moderately high levels who have tried and failed to lower their cholesterol enough with changes in lifestyle.

**Q:** But that diet and lifestyle stuff is hard. Why can't people just take drugs?

**A:** Because all the drugs used to lower cholesterol and triglycerides have potentially harmful side effects. Some have more, some have less, true, and some may have long-term side effects that are yet to be determined. But all experts now agree that drugs should be used to lower cholesterol only after a *determined effort* (their words, not ours) has failed to lower a person's LDL cholesterol enough to move her out of the high-risk-for-heart-disease category.

Also, drugs work better in combination with a low-fat diet, and doctors want people to take the lowest possible dose of a drug as they can. Experts say that for most people, a low-fat diet and treatment with a single drug, usually either a resin or a statin (we explain both below), works just fine.

**Q:** **How high does cholesterol need to be for a doctor to prescribe drugs?**

**A:** The general guidelines, set by the National Cholesterol Education Program, are these:

• For people who already have heart disease, an LDL cholesterol level of more than 130 mg/dL. (The goal is to reduce LDL to 100 mg/dL or lower.)

• For people without heart disease but two or more other risk factors, an LDL level greater than 160 mg/dL. (The goal is to reduce LDL to 130 mg/dL or lower.)

• For people without heart disease and fewer than two other risk factors, an LDL cholesterol level of 190 mg/dL. (The goal is to reduce LDL to 160 mg/dL or lower.)

**Q:** And what about for triglycerides?

**A:** High triglycerides are almost always found along with high LDL cholesterol. In that case, triglycerides need to be 200 mg/dL or higher for your doctor to consider using drugs.

If you have high triglycerides alone—a genetic condition—drug treatment is usually started at 500 mg/dL or so, depending on your other risk factors.

**Q:** Those recommendations sound like they are cut in stone. Are they?

**A:** Definitely not. In reality, the cutoff point at which drugs are clearly required is different for each person, so the decision whether to start drugs may be a complicated one. Doctors need to take the severity of other risk factors into consideration. Someone with severe diabetes or high blood pressure, for instance, is a more likely candidate for immediate drug therapy than someone who's slightly overweight and sedentary. And your age and whether you're a man or woman are always factors. That's where your doctor's individual counsel comes in.

**Q:** So not everyone needs to take drugs even though their LDL cholesterol levels remain a bit high?

**A:** Right. If you're not considered to be at high risk for developing heart disease over the next one or two decades, drug therapy often can be withheld or delayed even if your LDL cholesterol or triglycerides are higher than they should be. That may be the case for premenopausal women and men under age 35.

On the other hand, if you are considered to be at very high risk—if your LDL cholesterol level is well over 220 mg/dL, you have multiple risk factors or established heart disease—your doctor may want to start you on cholesterol- or triglyceride-lowering drugs shortly after starting dietary therapy. He's most likely to do that if he believes you won't be able to reduce your levels enough by dietary changes alone.

**Q:** So my doctor's beliefs about cholesterol-lowering diet and drugs might have something to do with what he recommends to me?

**A:** It might have a lot to do with it. More doctors than ever are prescribing cholesterol-lowering drugs, especially the newer, stronger drugs, **HMG CoA reductase inhibitors**, because of the findings of a study we described in Chapter 2, the so-called "4-S" study—the Scandinavian Simvastatin Survival Study.

In that study, a drug, simvastatin (Zocor), performed impressively. It reduced the overall risk of death by 30 percent and decreased the risk of death from heart attack by 42 percent. It also cut people's risk of having to undergo bypass surgery, and it reduced risks in women of any age and in people over age 60, two groups in which the benefits of cholesterol-lowering drug therapy have been less than certain.

Because of these impressive findings, the FDA has allowed Zocor's maker, Merck & Co., to relabel this drug as the first anticholesterol drug that actually reduces deaths from heart disease. And some doctors are urging heart patients to get their cholesterol levels rechecked and to consider trying this drug if their LDL cholesterol hasn't dropped to 100 md/dL by other treatments they may be using.

But some doctors believe the potential risks of even these new, apparently safer cholesterol-lowering drugs have been underplayed, and that new, long-term risks may yet emerge. These doctors are more reserved in prescribing. They believe cholesterol-lowering drugs should be used only in people with established heart disease whose cholesterol can't be controlled by diet alone.

And these doctors may be very aggressive about lowering your cholesterol and triglycerides with diet, and with lowering your other risks for heart disease with lifestyle changes, such as stopping smoking, losing weight and exercising.

**Q: Tell me again, how long am I supposed to be trying this diet and lifestyle stuff?**

**A:** At least three months, and doctors who are the most successful at getting results with diet say up to a year, even more if you are continuing to improve. As we said in Chapter 5, it takes time to incorporate healthier habits into your life. If your cholesterol is high and you're serious about avoiding drugs and heart disease, your best bet is to hook up early on with a dietitian or nutritionist to help you figure it all out.

## Q: What if I am already taking cholesterol-lowering drugs?

## A: They aren't necessarily a lifetime sentence. You may be able to reduce the dose or get off the drugs altogether (with medical supervision) if you can reduce your risks, including high cholesterol, with a renewed attempt at lifestyle changes.

## Q: What kinds of drugs are used to treat cholesterol?

## A: Basically, there are two types. Bile acid binding resins eliminate cholesterol from the body. These drugs were also mentioned in Chapter 4, since they are the only cholesterol-lowering drugs approved for use in children.

There are also drugs that block the liver's production of cholesterol. These include two major classes of drugs, the HMG CoA reductase inhibitors and the fibric acid derivatives. Also capable of blocking the production of cholesterol are a form of the B vitamin niacin (called nicotinic acid) and a drug in a class by itself, and not often used, called **probucol**.

—

# BILE ACID BINDING RESINS

**Q:** What are bile acid binding resins?

**A:** Well, for one thing, they are man-made resins. They are gritty, insoluble granules that come as a powder to be mixed with a liquid or as a bar that has to be chewed thoroughly. These resins bind with the cholesterol-rich bile acids secreted by the liver, preventing cholesterol from being reabsorbed into your body. Instead, it passes out of your body in feces. These drugs are also often called bile acid sequestrants, since they sequester, or seize, bile acid.

**Q:** How does that lower cholesterol in your blood?

**A:** When the normal recycling of bile acid back to the liver is inhibited, stores of cholesterol in the cells in the liver drop. The cells respond by drawing in more cholesterol from the blood to make bile acid. That results in a drop in both total and LDL cholesterol and a small increase in protective HDL cholesterol.

**Q:** How much of a drop can I expect to see using a bile acid binding resin?

**A:** Both of the bile acid binding resins available, cholestyramine (Questran, a powder) and colestipol (Cholybar, a chewable bar; and Colestid, granules) are equally effective in lowering LDL cholesterol.

People taking a normal daily dose can expect their LDL cholesterol to drop 15 to 25 percent, depending on how high it is before they start taking the drug.

# Q: What kinds of people take these drugs?

A: As we detail below, these drugs have a long safety record and seldom cause serious side effects. So they are a first choice for most people with high LDL cholesterol—men, women and sometimes children with cholesterol-elevating genetic disorders.

Some doctors also recommend bile acid resins for young adults, particularly men, with high LDL cholesterol levels but no other risk factors for heart disease, because of their increased odds for dying from heart disease later in life.

# Q: Are these drugs actually proven to reduce someone's chances of developing heart disease?

A: Yes. In the mid 1980s, they were shown to reduce coronary-heart-disease risk in the Lipid Research Clinics Coronary Primary Prevention trial.

# Q: What kinds of side effects do bile acid binding resins have?

A: These drugs are not absorbed into the body, so they have fewer side effects than other cholesterol-lowering drugs.

However, they can adversely affect the bowels. This drug makes stool dense and sticky, so two out of three people develop constipation, and some develop severe constipation and even fecal impaction—hardened or puttylike stools stuck in the rectum or colon.

How constipated someone gets depends on the dose. If he's taking only two scoops, packets or bars a day, constipation is less than if he's taking four or six doses. Constipation is also more likely, and more severe, in older people.

**Q:** **What can you do for the constipation? Take laxatives?**

**A:** Laxative pills are better avoided, since regular use can deplete the body of nutrients. But taking the water-soluble fiber **psyllium** (found in Metamucil, a bulk laxative) along with plenty of water was found to alleviate most of the symptoms of constipation in one study of people taking bile acid binding resins. An added benefit: Psyllium also removes some bile acids from the body.

Other complaints associated with this class of drugs include abdominal pain, heartburn, nausea, belching and bloating.

**Q:** Any other side effects?

**A:** The bile acid sequestering drugs can also interfere with the absorption of other medications you take, so other drugs should be taken at least one hour before the bile acid binding resin, or, if this is not possible, four hours afterward.

The drugs also interfere with the absorption of fat-soluble vitamins—A, D, E and K—and folic acid and iron. So if you're taking vitamin supplements, take them with a meal when you're not taking a bile acid sequestering drug or four hours after your last dose.

**Q:** Can these drugs cause vitamin deficiencies?

**A:** At the usual dosages these drugs are given, they are not known to cause vitamin- or mineral-deficiency-related problems, such as anemia. Vitamin deficiencies can be hard to detect, however—especially for fat-soluble vitamins—and doctors don't often even think to look for them. So keep a lookout yourself. Signs of deficiency of these nutrients may include depression, easy bruising or bleeding (including nose-bleeds), fatigue, weakened immunity, dry eyes, night blindness and bone pain.

**Q:** Any other side effects I should know about?

**A:** Sometimes blood triglycerides rise when you take a bile acid binding resin. The increase may be transient, but it persists in some people and may require the use of a second lipid-lowering drug, such as nicotinic acid, which we describe next, to lower triglyceride levels.

# NIACIN (NICOTINIC ACID)

**Q:** I've heard that a vitamin, niacin, is used to lower cholesterol. What can you tell me about it?

**A:** Only one form of niacin, nicotinic acid, can lower cholesterol. Another form, nicotinamide, has no effect on cholesterol.

**Q:** Nicotinic acid. Isn't that found in cigarettes?

**A:** No. Despite similarities in names, nicotinic acid is not related to nicotine, the addictive substance found in tobacco.

**Q:** How do I get nicotinic acid? In foods?

**A:** While there's some in foods, there's nowhere near enough to lower your cholesterol. Most people get only 15 to 35 milligrams of niacin a day from foods. At least 60 times that amount is required (as nicotinic acid) to help lower cholesterol. The usual dosage is 1½ to 3 grams a day, although as much as 6 grams a day has been prescribed. You can get nicotinic acid without a prescription, but it's not wise to use it without medical supervision, as we detail later.

# Q: How does it work?

A: Its main action is in the liver, where it interferes with cholesterol manufacture, probably in several ways, through mechanisms that are not well understood. Its effect on the liver leads to decreased production of triglyceride-rich VLDL cholesterol. And since VLDL cholesterol is eventually converted to LDL cholesterol, nicotinic acid reduces both LDL cholesterol and triglyceride levels.

# Q: How much of a reduction can I expect to see?

A: The higher your cholesterol to begin with, the more of a drop you'll see. People with total cholesterol of 240 mg/dL or higher might see a 10 to 20 percent drop in total cholesterol and in LDL cholesterol levels, a reduction comparable to that achieved with normal doses of 20 to 40 milligrams a day of lovastatin, a drug we talk about in a minute.

Nicotinic acid also reduces blood levels of two proteins thought to be linked with heart disease: homocysteine and methylmalonic acid, which are also reduced by folic acid.

# Q: What about HDL cholesterol? Does nicotinic acid affect it?

A: Yes. Nicotinic acid has the greatest HDL-raising effect of any drug. In gram doses, it increases HDL cholesterol levels an average of 20 to 35 percent.

That's good, because studies suggest that a 1 mg/dL increase in HDL cholesterol levels should decrease coronary-heart-disease risk by 2 to 3 percent. By reducing LDL and increasing HDL cholesterol levels, nicotinic acid can improve people's HDL-to-total-cholesterol ratio, a figure that some doctors think is the best indicator of your risk for developing heart disease.

**Q: So should I take nicotinic acid if my total cholesterol is normal but my HDLs are low?**

**A:** Right now the National Cholesterol Education Program recommends against drug therapy for people with normal total or LDL cholesterol levels but low HDLs. Exercise can raise HDLs. We talk more about exercise in Chapter 7.

**Q: Is nicotinic acid proven to reduce the risk for heart disease?**

**A:** Yes. It has a fairly impressive history. One study, the Coronary Drug Project, suggested that nicotinic acid can protect against heart disease both while it's being used and long after it's stopped, something no other drug has been proved to do. In a follow-up 10 years after that five-year study ended, those who had taken 2 grams a day of nicotinic acid had 20 percent less coronary heart disease than a corresponding group that had taken a placebo.

## Q: Now give me the bad news. What are the side effects?

A: Nicotinic acid has been around for a while. It was first found to lower cholesterol back in 1955, and it's been used extensively for at least 20 years, so most of its side effects are known.

The most common side effect is something called the niacin flush. About half an hour after you take it, you'll feel warmth, flushing and itchiness, usually on your face and upper body. This effect lasts only 15 to 30 minutes.

## Q: Any way to get around this?

A: Doctors say that, in most people, the flush is worst when they first start taking the drug—one good reason to start it on a weekend—and that it soon lessens as the body adapts to the drug.

And the flush can be greatly minimized if you gradually increase your dose of nicotinic acid, starting with as little as 100 milligrams, three times a day, slowly increasing the dose until you reach your initial target level, usually 500 milligrams, three times a day. Taking nicotinic acid with a meal also helps. So does taking one regular aspirin with the nicotinic acid. Avoid taking the drug with alcohol or hot drinks, which usually makes the flush worse.

**Q:** Aren't there more serious side effects associated with nicotinic acid? I heard it can cause liver problems.

**A:** It can. A regular dose can cause problems, but the timed-release form, which causes less flushing, is even more likely to cause this potentially serious side effect. Only the timed-release form, however in amounts of 500 milligrams or more a day, has been reported to cause severe, permanent liver damage. Most doctors don't recommend timed-release nicotinic acid for this reason.

**Q:** How can nicotinic acid cause liver damage?

**A:** In the large doses used to lower cholesterol, nicotinic acid can irritate liver cells. If it's going to do this, it usually does soon after you start the drug. A doctor can check for liver damage by testing your blood for certain enzymes released by the liver—a liver function test. This test should be done four to six weeks after you have reached your initial target dose, usually 1½ grams of niacin, and then, on a regular basis.

**Q:** What's a typical dose? 1½ grams a day?

**A:** That's an amount plenty of people stick with, although doubling that amount to 3 grams a day usually produces additional good results. More than 3 grams, however, and you begin to see diminishing returns. Your cholesterol just doesn't drop that much lower.

**Q:** Does the regular form ever cause permanent liver damage?

**A:** Usually not. The changes in liver function usually last only a few weeks and are moderate in degree. But sometimes people do have a severe reaction to nicotinic acid and have to stop taking it. Even then, their livers usually recover in two to three months. Occasionally people start taking nicotinic acid again and have no problems with it. Still, they need to be monitored carefully.

**Q:** So I guess people with liver problems shouldn't take nicotinic acid?

**A:** It's generally off limits to people with severe or unexplained liver problems, and should be used only with caution in people with a history of stomach or intestinal ulcers. If you have an active ulcer, you shouldn't be taking it at all.

If you tend to have gout attacks, nicotinic acid may not be the best cholesterol-lowering drug for you, since it can increase blood levels of uric acid and bring on an attack of gout. During an attack, blood levels of uric acid get so high that some of it crystallizes in joints, causing acute pain.

And people with diabetes must be carefully monitored if they take this drug, since it can make their blood-sugar levels harder to control.

**Q:** I've heard that, pricewise, niacin is the least expensive cholesterol-lowering drug around. Is that true?

**A:** Yes. It's definitely a bargain compared with other cholesterol-lowering drugs. A 10-day supply of 500-milligram tablets costs about $2. Prescription brands (Niacor, Nicolar) are more expensive, so you may want to stick with generic brands.

**Q:** Do you need a doctor's prescription to take nicotinic acid?

**A:** No. Nicotinic acid can be obtained without prescription at a pharmacy or health-food store. But because of the possible side effects we discussed, it's important not to take large amounts of nicotinic acid without medical supervision.

## STATIN DRUGS: HMG CoA REDUCTASE INHIBITORS

**Q:** I've heard there are some strong new drugs that can really lower cholesterol? What are they?

**A:** You're referring to a class of drugs called "the statins" because their chemical names all end in "statin." These drugs are officially called HMG CoA reductase inhibitors, and they include lovastatin

(Mevacor), the oldest of these drugs, marketed since 1982, and three closely related newcomers—pravastatin (Pravachol), simvastatin (Zocor) and fluvastatin (Lescol).

**Q:** How do these drugs work?

**A:** Like nicotinic acid, drugs in this class work in the liver to block the manufacture of cholesterol. They do that by inhibiting the action of a key enzyme involved in cholesterol production, called HMG CoA reductase. As the amount of cholesterol being produced in the liver drops, the cells in the liver take up more LDL cholesterol from the blood, so blood levels of LDL cholesterol drop.

**Q:** How much of a drop in cholesterol levels can I expect to see with these drugs?

**A:** That depends on your dosage and how high your cholesterol is to begin with. Generally, though, studies show a 20 to 30 percent drop in both total and LDL cholesterol levels. Triglycerides also drop with these drugs, from 7 to 20 percent. And HDL cholesterol increases, from 6 to 12 percent.

**Q:** How's that compare with the other drugs you already discussed?

**A:** All indications are that the HMG CoA reductase inhibitors work at least as well as the bile acid binding resins and nicotinic acid in preventing coronary heart disease.

It was a statin drug (simvastatin, used in the "4-S" study), you'll recall, that caused a reduced risk of 30 percent from death from any cause and a reduced risk of 42 percent from heart attack.

**Q:** Is simvastatin better than the other statin drugs?

**A:** No tests have been done to detect differences in ability to lower cholesterol among these drugs, presumably because the pharmaceutical companies believe they are pretty much similar. But most of the work that's been done on these drugs indicates that they work equally well at lowering cholesterol.

On a milligram-per-milligram basis, simvastatin appears to be more potent than the other two, but equivalent cholesterol reductions can be achieved with the other agents if higher doses are used.

**Q:** What about side effects? Are they the same in all these drugs?

**A:** They seem to be, although since lovastatin has been around the longest, more is known about its long-term side effects than for the other drugs.

# Q: What are the side effects?

A: They are numerous, they can be serious and some doctors believe not all the long-term side effects of these drugs have been ferreted out yet.

Out of every 100 people who take these drugs, about two will have liver-function problems. So everyone taking these drugs needs to undergo liver-function tests about four to six weeks after they start taking the drugs, and then twice a year.

Muscle inflammation is also a serious side effect, reported in about 1 in 100 people taking these drugs. Its incidence increases when HMG CoA reductase inhibitors are combined with other drugs, including gemfibrozil, nicotinic acid, cyclosporin and erythromycin. So you'll need to avoid these drugs if possible.

Since this problem can be associated with the breakdown of muscle tissue and, along with it, kidney failure, it's vitally important to tell your doctor if you develop weak or tender muscles, backache or what seems like "rheumatism" or excess muscle stiffness after exercise.

Blood tests that measure by-products of muscle breakdown should be done regularly to alert your doctor to early signs of inflammation.

# Q: Any other side effects?

A: Constipation, diarrhea, flatulence and nausea can occur, although these symptoms are less common than with the bile acid binding resins. Headaches, dizziness and blurred vision are also possible.

In animals taking much higher doses of the drugs than would normally be given to humans, lovastatin has produced optic-nerve degeneration and lesions in the central nervous system. However, these nerve problems have not been noted in people taking the drug.

**Q:** Is it possible I shouldn't take one of these drugs?

**A:** Sure. Since there are so many potential side effects, you should think twice about using any of these drugs. You may be able to reduce your cholesterol sufficiently using diet, bile acid binding resins, and perhaps some nicotinic acid. If your cholesterol is in the 200 to 240 mg/dL range, you should weigh the risks of taking these drugs with their potential benefits.

The manufacturers say these drugs should be "prescribed with caution" to people with liver disease or possible liver or kidney problems, women who are breast-feeding their children, pregnant women, and women of childbearing age unless they are highly unlikely to become pregnant.

## FIBRIC ACID DERIVATIVES

**Q:** What are fibric acid derivatives?

**A:** These are cholesterol- and triglyceride-lowering drugs that have been around since the late 1960s. They were initially prescribed with some enthusiasm because they seemed to have fewer side effects than bile acid resins and nicotinic acid—the two other

serious contenders at the time—but they were found
to have their own very serious side effects. So today
they are prescribed only for certain conditions, which
we'll describe soon.

**Q:** What are the names of these drugs?

**A:** Two types of fibric acid derivatives are currently
available in the United States: gemfibrozil (Lopid)
and clofibrate (Atromid-S).

**Q:** How do fibric acid derivatives lower
cholesterol?

**A:** They both work in the liver, but the way they
work is not well understood. They increase the
liver's ability to break down VLDL cholesterol, which
drops triglyceride levels. And they raise HDL cholesterol
levels slightly.

**Q:** So what is the problem with these drugs?

**A:** It's true that both of these drugs have been
shown to reduce the risk of fatal and nonfatal
heart attacks in two large studies of people with no
signs of heart disease. However, there was an increase
in total deaths in people taking clofibrate in one of
these studies, the World Health Organization trial.

In the other study, the Helsinki Heart Study, gemfibrozil did not cause an increase in deaths during the study, but in an 8½ year follow-up, the people who had taken gemfibrozil during the study had a 20 percent increased rate of death compared with the placebo group.

As a result of these findings, prescriptions of clofibrate and its closely related cousin, gemfibrozil, have dropped off considerably, with good reason.

## Q: What happens to people who take these drugs?

**A:** With clofibrate, half the deaths were due to malignancies, such as liver cancer. Some were due to gallbladder disease or complications from gallbladder surgery. Other studies have reported an increase in heart arrhythmia, blood clotting and angina in people taking clofibrate.

## Q: And with gemfibrozil?

**A:** Because it's so closely related to clofibrate, it's considered to have the same potential risk for toxicity, including cancer and gallbladder problems, plus an increased risk for deaths not linked to heart disease.

**Q:** What about less serious side effects?

**A:** The list is long: Stomach upset, nausea, diarrhea, rash, muscle pain, weakness, liver-function problems, dizziness and blurred vision are just some potential side effects.


**Q:** So who are these drugs prescribed for these days—masochists?

**A:** Clofibrate or gemfibrozil may be prescribed for people who have very high triglyceride levels—2,000 mg/dL or more—as the result of genetic disorders. These people may also be at high risk of developing pancreatitis, a serious, painful inflammation of the pancreas.

Gemfibrozil is prescribed more often than clofibrate, and may also be used to prevent the development of heart disease in people with a combination of high LDL cholesterol, high triglycerides and low HDL cholesterol. However, benefits from the drug seem to be limited to people with LDL/HDL ratios of more than 5:1 and triglycerides higher than 200 mg/dL.


**Q:** How much do these drugs reduce triglycerides?

**A:** They do a pretty good job, reducing triglyc-erides 20 to 50 percent while increasing HDL cholesterol 10 to 15 percent. But they don't lower LDL cholesterol very much—only 10 to 15 percent.

## PROBUCOL

**Q:** What is probucol?

**A:** Probucol (brand name Lorelco) is a cholesterol-lowering drug that's been on the market since 1977. Its chemical structure is not similar to any other class of cholesterol-lowering drugs.

Basically, it is a drug of last resort and one you'll want to avoid taking in most situations. Its use is restricted to people who have not tolerated or responded to other cholesterol-lowering drugs.

**Q:** Why shouldn't I take this drug?

**A:** Lorelco does reduce LDL cholesterol somewhat, by 10 to 20 percent, but it also decreases HDL cholesterol 10 to 30 percent. And it does not lower triglycerides.

**Q:** So it doesn't work very well compared with some other cholesterol-lowering drugs. What about side effects?

**A:** Although most people do alright on this drug, it does have occasional serious side effects. It can cause irregular heartbeat and syncope (a temporary loss of consciousness due to lack of blood to the brain). Sudden death is considered one of the potential side effects of this drug, along with gastrointestinal bleeding, impotency, insomnia, nerve damage, ringing ears and a long list of other symptoms.

**Q:** **Do doctors ever prescribe combinations of cholesterol drugs? Is this more dangerous than using only one?**

**A:** Drugs sometimes are combined to treat very high cholesterol or in an attempt to keep dosages low. Some combinations give good results.

For instance, either nicotinic acid or statin drugs can be used along with bile acid binding resins without added side effects. In the case of statin drugs, this may allow someone to take the smallest dosage of this drug, 20 milligrams a day. Used alone, that would result in an average drop in LDL cholesterol of 24 percent. But used in combination with the lowest dose of a bile acid binding resin, LDL cholesterol drops up to 43 percent. Of course, a patient needs to take each of these drugs separately during the day, since the bile acid binding resin can interfere with the body's ability to absorb the statin drug.

Some combinations can be dangerous, however, and are not used. Combining nicotinic acid and a statin drug, or two statin drugs, for instance, may lead to an increased risk for serious muscle inflammation.

# Q: How do I know if my doctor is prescribing the right drug, or drugs, for me?

**A:** Ask your doctor why she is prescribing a particular drug for you. Is it because your triglycerides are particularly high? Because your HDLs are low? Because your LDLs are still way too high after a year or more of changing your lifestyle and diet?

Ask her what she expects the drug to do and how she expects to monitor you for side effects. Make sure your doctor knows of any other drugs you are taking, prescription or nonprescription. Ask her about any possible interactions. You want to make sure your doctor is making a decision based on your individual needs, not on the last drug-company salesperson who was in her office.

When you go to have your prescription filled, ask the pharmacist for an information sheet on your drug. Make sure you get the real thing, not an abbreviated patient information sheet. Read it and keep it on hand to refer to later, if necessary, for possible side effects.

This information is also available in the *Physician's Desk Reference (PDR)*. Most libraries have a copy of this book in their reference sections.

# 7 OTHER FACTORS THAT AFFECT CHOLESTEROL AND HEART DISEASE

**Q:** You mentioned some other risk factors for heart disease—smoking, stress, lack of exercise and the like. Why are these risk factors? Do they raise cholesterol levels?

**A:** All of them can contribute to heart disease in a number of ways: by causing the oxidation of LDL cholesterol, making blood quicker to clot, damaging artery linings, causing blood vessel-constriction, and raising cholesterol levels or lowering HDL cholesterol.

## SMOKING

**Q:** I know smoking is bad for the heart. But how bad? And why?

**A:** Cigarette smoking is an important risk factor for heart disease, for everyone—men and women—of any age. Heart attacks are 2½ times more common in smokers than in nonsmokers. A comparable increase in risk would come from having a cholesterol level of 300 mg/dL or more. Smoking also takes away a woman's premenopausal protection from estrogen, the main female hormone, since smoking depletes estrogen in the body.

**Q:** But what if I don't smoke all that much?

**A:** It's true that the risk increases with the number of cigarettes you smoke each day. But if you smoke even a pack a day, your risk for heart disease is twice as high as that of someone who never smoked. That risk is comparable to a total cholesterol level of 300 mg/dL.

If you smoke two or more packs a day, your risk is three times as high as a person who never smoked. Cigars, pipes and chewing tobacco also increase your risk—perhaps as much as two times as high as a person who never smoked. And the earlier in life you start smoking, the more negative impact it's likely to have on your health.

**Q:** Does smoking low-tar and low-nicotine cigarettes reduce my risk for heart disease?

**A:** Sorry, it doesn't. While these cigarettes may reduce your risk for cancer, they have not been proved to reduce your risk for heart disease. Studies show that people who smoke low-tar and low-nicotine cigarettes often inhale more deeply, hold their breath after inhaling, and smoke more cigarettes in an unconscious effort to maintain the nicotine levels to which their bodies are addicted. Consequently, they not only do not reduce their nicotine exposure as much as they may have hoped but also inhale more of the other toxic substances contained in the smoke.

**Q:** How exactly does smoking affect the heart?

**A:** Smoking causes the oxidation of lipids, the first step in the process of plaque buildup. It reduces the ratio of HDL to LDL cholesterol and increases the tendency for blood to clot inside the blood vessels and obstruct blood flow. Constituents of tobacco smoke also directly damage endothelial cells, the cells lining the inside of blood vessels.

**Q:** Sounds bad.

**A:** It is. And cigarette smoking also has temporary adverse effects on the heart and blood vessels, and these may provoke serious consequences, such as heart attacks. The nicotine in smoke increases blood pressure and heart rate. Carbon monoxide, a gas produced by smoking—the same gas in car exhaust that is lethal in an enclosed space—gets into the blood and reduces the amount of oxygen blood can carry to the heart and the rest of the body. It causes arteries in arms and legs to constrict, and can cause a lack of oxygen and blood flow to the heart muscle by temporarily decreasing the diameter of the coronary artery.

**Q:** If someone quits smoking, how long will it take for her risk of heart disease to drop?

**A:** Not long. A major reduction in risk occurs within the first year after she quits, and the risk drops dramatically within about two years. Although the risk never drops to where it would be had she never smoked, it will eventually drop so that it's only about one-third higher.

**Q:** What if I've been smoking for years? Does it really help to stop?

**A:** Yes! For example, if you are older than 50 and stop smoking, your chance of dying from any cause is reduced by one-half during the next 15 years. And your risk of dying of heart disease decreases by about 30 percent.

**Q:** I don't smoke, but I'm exposed to cigarette smoke all day long at work. Does that increase my risk for heart disease?

**A:** Yes. Studies find that exposure to cigarette smoke in the workplace or at home increases your risk for heart disease from 20 to 70 percent. In fact, passive smoking has a stronger link with heart disease than it does with lung cancer. Environmental experts predict that each year nearly 40,000 Americans die of heart disease aggravated by passive smoking.

## EXERCISE

**Q:** **I've heard exercise is supposed to help prevent heart disease? Does it?**

**A:** Yes. Sedentary people have nearly twice the risk of having a fatal heart attack as active people when other factors are equal.

**Q:** **What's it mean to be sedentary?**

**A:** You are sedentary if you have a job that is inactive or spend most of your day sitting and do not take time to exercise 20 to 30 minutes at least three times a week.

**Q:** **Is that 20 to 30 minutes three times a week what's required to protect your heart?**

**A:** Not necessarily. It's simply enough to move you out of the "sofa spud" category. Studies suggest that the best protection comes getting four or five hours a week of aerobic exercise—the huff-and-puff kind.

**Q:** How does exercise protect your heart? Does it lower cholesterol?

**A:** Regular exercise raises blood levels of "good" HDL cholesterol. How much it goes up depends on your initial cholesterol level, age, weight and amount of body fat, as well as the intensity of your workouts.

**Q:** How much of an increase can someone expect to see?

**A:** Endurance athletes, such as long-distance runners or swimmers, often have HDL levels 10 to 24 mg/dL higher than those of people who don't exercise at all. And studies also indicate that moderate exercise, such as walking briskly for an hour three times a week, also raises HDLs.

**Q:** Does exercise do anything else to help prevent heart disease?

**A:** If it helps you lose weight, you'll also probably have a drop in LDL cholesterol and triglycerides and you may reduce your risk of developing diabetes or high blood pressure, two other big risk factors for heart disease.

Aerobic exercise strengthens your heart muscle, making it pump more blood with each beat. Exercise also makes the platelets in your blood less sticky, which helps to reduce the possibility of blood clotting in arteries. And in animals, at least, regular exercise stimulates the formation of collateral coronary arteries, providing blood supply to the heart by going around blockages. Whether that also happens in humans is not known.

**Q:** Sounds like everyone should exercise. But don't people who already have heart disease drop dead when they start running?

**A:** It's true that exercise can precipitate sudden cardiac death, especially in people with known heart disease.

But according to two studies, such cases are rare, even though they often make headlines. Most at risk are so-called weekend warriors, people who throw themselves into strenuous activities without first taking the time to get into shape. People who exercise regularly also increase their risk of dying during peak exertion, but their overall risk of sudden death is actually 60 percent lower than in people who never exercise. So the exercise really is protecting their hearts.

## STRESS

**Q:** Is it true that a person can get so stressed out that he has a heart attack?

**A:** Yes. It's not uncommon for people with heart disease to report that emotional peaks cause chest pain. And it's common for heart attacks to occur during emotionally difficult times. However, most people who have stress-related heart symptoms have underlying atherosclerosis.

**Q:** Can stress cause the development of heart disease?

**A:** That's not entirely clear, in part because stress —and people's reactions to it—are so hard to study and measure. Some doctors, like Dean Ornish, M.D., director of the Preventive Medicine Research Institute, in Sausalito, California, believe that stress plays a major role in the development of heart disease and heart attacks.

One recent study, for instance, found that men who had rated themselves highly anxious back in 1961 were three to four times more likely to go on to have a sudden fatal heart attack than men who rated themselves low in anxiety.

**Q:** Does stress raise cholesterol levels?

**A:** Some doctors claim no. But a recent study by researchers at the University of Pittsburgh found just the opposite. They determined that within 20 minutes of starting on a frustrating mental task people had an increase of up to 5 mg/dL in total cholesterol, with a slightly lower rise in triglycerides, HDL and LDL cholesterol.

**Q:** Can't stress also hurt my heart in other ways?

**A:** Apparently so. The biochemicals your body releases during stress trigger processes that can harm the artery walls. They can also make blood thicker and more prone to clotting.

Stress raises blood pressure and blood-sugar levels, speeds up the rate of heartbeat, contributes to heartbeat irregularities (arrhythmia) and causes blood vessels to constrict. It can even produce coronary spasms, thus narrowing arteries and reducing the flow of blood to the heart. Mental stress, especially anger, has been reported to trigger angina, heart attacks, even sudden death.

**Q:** What can you do if you're under stress or overreact to things?

**A:** For starters, take good care of yourself. Don't smoke or stuff yourself on fatty foods. Exercise is a great way to reduce stress. It actually uses up the "fight or flight" biochemicals produced in a stressful episode epinephrine and norepinephrine, adrenaline-like chemicals.

Yoga, meditation, spiritual pursuits, learned relaxation techniques, talking it out with a friend or therapist, social activities and just plain having fun are good ways to diffuse stress. People with heart disease are also much more likely than normal to be depressed, and they can often benefit from drugs used to treat depression.

## OVERWEIGHT

**Q:** Does being overweight raise cholesterol levels?

**A:** It can. People who are obese (more than 20 percent overweight) do tend to have higher LDL and lower HDL cholesterol than normal-weight people. But excess body fat doesn't necessarily cause these cholesterol changes. It may simply be found along with them. People who are obese also tend to have other risk factors for heart disease: high triglyceride levels, high blood sugar, high blood pressure, diabetes.

**Q:** So does being obese increase my chances of having a heart attack?

**A:** Study findings are mixed on this. Because obesity is so often found along with the other risk factors we just mentioned, it's hard to know if obesity, by itself, is a risk for heart disease. Some research seems to suggest that if your only risk is being fat—no high blood pressure or cholesterol, no diabetes—you're probably not at increased risk. Other studies find that even a slight increase in weight—10 to 15 pounds—ups your risk. Take your choice.

**Q:** I've heard that having a potbelly increases risk for heart disease more than being fat all over. Is that true?

**A:** Yes. Many women have fat on their hips and thighs that is not associated with increased risk, whereas potbellies (visceral fat)—on men or women— are linked with heart disease. People with potbellies often also have high insulin and high blood-sugar levels, which also contribute to heart disease.

**Q:** Does losing weight help to reduce my risk for heart disease?

**A:** It does if you have other risk factors associated with obesity. If you can lose weight—and keep it off—you may be able to eliminate all the risk factors we just mentioned. Losing weight is often the only thing that's needed to reverse diabetes and high tri- glycerides. And losing just 5 or 10 pounds can double the reduction you'll see in LDL cholesterol on a low- saturated-fat diet.

**Q:** Do thin people ever have high cholesterol or atherosclerosis?

**A:** Sure. While overweight people are about twice as likely as thin people to have high cholesterol, thin people can also have high cholesterol. Usually, it is the result of an inherited abnormality in cholesterol metabolism, which can make their high cholesterol more difficult than normal to treat.

## RACE

**Q:** **What about race? Are white people more vulnerable to heart disease than black people?**

**A:** No, the opposite is true. Heart-disease death rates are 3 to 70 percent higher among blacks than among whites of the same age, at least up until age 74, when it evens out. The reason for this difference is not entirely clear, but it apparently does not have to do with higher cholesterol levels, since HDL cholesterol is actually higher in blacks than in whites.

It may be that black people are more likely than white people to have other additional risk factors. They continue to smoke more than whites, for instance, and a higher percentage have high blood pressure or diabetes.

**Q:** **Wow! Heart disease—even just one aspect of it, cholesterol—is more complicated than I ever thought. How am I supposed to keep all this stuff straight, much less do anything about it?**

**A:** Don't despair. Just keep in mind these points:

• Know your numbers. But remember, high cholesterol is not heart disease; it is not even a guarantee that you'll get heart disease. It is simply a measure of risk. Make sure your cholesterol is measured accurately; ask your doctor to tell you specifically how your numbers affect your risk for heart disease. Ask for copies of your tests and of your medical records, which should also include other "numbers" you need to know: blood pressure, blood sugar and your weight.

• Know your other risk factors for heart disease and be prepared to change the top two or three. Persist in getting the help you need. Ask your doctor about help in stopping smoking, beginning exercise or dieting. She can prescribe nicotine gum or patches and refer you to a psychologist or hypnotist for additional help. Such professionals may also be able to help you devise tactics for stress reduction and weight control.

Even a one-time referral to a physical therapist can help you get started on an exercise program. Or if you already have heart disease, ask about enrolling in a cardiac-rehabilitation program. Call around to local hospitals to find out what they offer in the way of heart-disease-prevention programs ("wellness" programs, in some cases). At your favorite health-food stores, get leads to alternative health-care professionals. (But check them out just as you would any health-care provider.)

• Improve your diet if necessary. See a dietitian or nutritionist for help, and refer to books on the subject by experts. (See "Suggested Reading" at the back of this book.)

• Stay as up-to-date as you can by continuing to read about heart disease and cholesterol. If you read about a particularly interesting study in the newspaper, go to the nearest hospital library and read the original study, usually published in a professional journal. And ask your doctor what she thinks about it the next time you see her.

# INFORMATIONAL
# AND
# MUTUAL-AID GROUPS

**American Association of Naturopathic Physicians**
2366 Eastlake Ave. East
Seattle, WA 98102
206-323-7610

> *These doctors take a "whole body" approach to heart*
> *disease and high cholesterol, and are specifically*
> *trained in nutrition and herbal therapies. Write for*
> *a list of naturopathic doctors in your area.*

**American Dietetic Association**
430 Michigan Ave.
Chicago, IL 60611

> *For a list of registered dietitians in your area, send*
> *your request, along with a self-addressed stamped*
> *envelope and $1. Or call the National Center for*
> *Nutrition and Dietetics' Consumer Nutrition*
> *Hotline at 800-366-1655.*

**American Heart Association**

> *Contact your local affiliate or call 800-AHA-USA1*
> *(800-242-8721) for a list of publications, including*
> *cookbooks and brochures.*

## American Holistic Medical Association
2002 Eastlake Ave. East
Seattle, WA 98102
206-322-6842

> *These doctors, too, are likely to take a "whole body" approach to heart disease and high cholesterol. Can refer you to a doctor in your area who is a member of this organization.*

## National Cholesterol Education Program
NHLBI Information Center
P.O. Box 30105
Bethesda, MD 20824-0105

> *Offers publications to help you lower your blood cholesterol.* One of these, So You Have High Blood Cholesterol, *gives detailed information on high blood cholesterol and how it affects your health.* Another, Step by Step: Eating to Lower Your High Blood Cholesterol, *gives details on cholesterol-lowering diets. Write for a list of publications.*

# GLOSSARY

**Absolute risk:** Actual risk; your "odds," as in 1 in 100.

**Angina pectoris:** Chest pain, often with an accompanying feeling of suffocation, caused by insufficient oxygen to the heart muscle.

**Antioxidant:** A substance with the ability to interfere with oxygen-generated, or oxidative, reactions. LDL cholesterol, for instance, undergoes oxidation as part of the early stages of atherosclerosis.

**Apolipoproteins:** Any of the protein constituents of lipoproteins; some apolipoproteins may prove to be markers for inherited cholesterol disorders.

**Arteriography:** A test that can measure obstruction inside an artery.

**Arteriosclerosis:** A broad term used to cover a variety of diseases, including atherosclerosis, that lead to abnormal thickening and hardening of the walls of the arteries.

**Atherosclerosis:** The gradual buildup of fatty deposits, called plaque, on the inside walls of the arteries.

**Attributable risk:** The amount of risk that can be pinned on a specific risk factor, such as smoking.

**Beta-carotene:** An orange pigment, found in vegetables and fruits and that acts as an antioxidant in the body.

**Bile acid binding resins:** Cholesterol-lowering drugs that act by binding with cholesterol-laden bile in the intestines, making it unabsorbable.

**Calcification:** The gathering of calcium deposits in body tissue. These deposits harden arteries.

**Canola oil:** A bland, light-colored vegetable oil that is mostly monounsaturated fats. These fats help drop LDL cholesterol levels.

**Carotid ultrasound test:** A noninvasive test that uses sound waves to measure the extent of calcification (hardening) in the carotid arteries of the neck.

**Cerebrovascular disease:** Blockages or disruption of blood circulation in the brain, including stroke.

**Cholesterol:** A white, waxy substance found naturally throughout the body, belonging to a class of compounds called sterols.

**Cholestyramine:** A cholesterol-lowering drug; a bile acid binding resin.

**Chylomicrons:** Tiny spheres of fat, mostly triglycerides, that go directly from the intestines to the liver and that are thought to have no direct effect on the development of **atherosclerosis**.

**Clofibrate (Atromid-S):** A drug that lowers serum cholesterol by reducing **very-low-density lipoprotein**.

**Colestipol:** A cholesterol-lowering drug; a bile acid binding resin.

**Complete lipid profile:** A blood test that measures total cholesterol, HDL, triglycerides and, indirectly, LDL cholesterol.

**Coronary-artery disease:** Atherosclerosis (blockage) of the coronary arteries, the spaghetti-size arteries that deliver blood to the muscles of the heart.

**Electrocardiography:** A test that measures and records electrical activity in the heart.

**Endothelial cells:** A layer of cells that lines and protects inner surfaces of blood vessels.

**Familial combined hyperlipidemia (FCH):** An inherited disorder of both high cholesterol and triglycerides; a serious condition that affects about 1 percent of the population and, untreated, results in premature heart disease.

**Familial hypercholestemia (FH):** The most common form of inherited high cholesterol, caused by a defect in LDL receptors, the portals on cell membranes that allow cholesterol to move in and out.

**Fibric acid derivatives:** A class of drugs that interfere with the body's ability to make cholesterol in the liver; includes **gemfibrozil** and **clofibrate**.

**Foam cell:** A large, foamy-looking type of immune cell—a macrophage—after it has eaten lots of cholesterol. Foam cells contribute to the development of atherosclerotic plaque.

**Gemfibrozil (Lopid):** A lipid-regulating drug that decreases serum triglycerides and **very-low-density lipoprotein**.

**High-density lipoprotein (HDL):** So-called good cholesterol that helps to escort cholesterol from the body. High levels are linked with reduced risk for heart disease.

**HMG CoA reductase inhibitors:** A class of drugs that interfere with the body's ability to make cholesterol in the liver; includes the statin drugs.

**Hydrogenated fats:** Oils processed so that hydrogen is added to their structure. This hardens the oil, but also makes it similar to saturated fats. Margarine and vegetable shortenings are hydrogenated fats.

**LDL receptors:** Portals on cell membranes that selectively allow LDL cholesterol to move in or out of a cell.

**Linoleic acid:** One of two essential fatty acids that cannot be manufactured in the body, found only in vegetable oils such as canola, safflower and corn oil.

**Linolenic acid:** An essential fatty acid, polyunsaturated, found in soybean, canola and nut oils.

**Lipid:** A group of fats or fatty substances found in the body.

**Lipoprotein:** A combination of lipids and protein.

**Low-density lipoprotein (LDL):** So-called bad cholesterol. High levels of LDL cholesterol have been linked with increased risk for heart disease.

**Macrophage:** An immune cell that engulfs and consumes microorganisms and debris, including oxidized cholesterol.

**Monounsaturated fats:** Fats, such as olive or canola oil, that are only slightly saturated. Diets rich in monounsaturated fats have been linked with reduced levels of LDL cholesterol.

**Niacin:** One of the B vitamins. One form of niacin, nicotinic acid, is used to lower cholesterol.

**Nicotinic acid:** A form of niacin, a B vitamin, used to lower cholesterol.

**Occlusive peripheral vascular disease:** Blockages of blood circulation in the legs or arms, including intermittent claudication, a painful condition caused by blockage of arteries in the legs.

**Omega-3 fatty acids:** A form of fat, found in fatty fish and some plants, that may help to prevent heart disease.

**Oxidation:** A chemical reaction that involves oxygen. LDL cholesterol becomes oxidized in the body in an early stage of atherosclerosis. Vitamin E and other antioxidant nutrients, such as vitamin C and beta-carotene, can prevent oxidation.

**Palmitic acid:** The predominant saturated fat in red meat, butter fat, cottonseed oil and cocoa butter; it raises LDL cholesterol.

**Pancreatitis:** Inflammation of the pancreas, often aggravated by superhigh levels of triglycerides.

**Plaque:** A patch or flat area where cholesterol has been deposited on an artery wall. Fatty plaques start the process of atherosclerosis.

**Platelet:** A small, disk-shaped structure involved in blood coagulation.

**Platelet aggregation:** A process in which platelets stick together and stick to artery walls; a part of blood clotting.

**Polyunsaturated fat:** The predominant fat in common vegetable oils like corn, safflower, sunflower, cottonseed, soybean and walnut oils.

**Probucol:** A cholesterol-lowering drug (brand name Lorelco) that is used only in people who can't tolerate other cholesterol-lowering drugs.

**Progesterone:** A female hormone; hormone-replacement therapy containing progesterone appears to raise triglycerides less significantly than therapies which do not contain it.

**Psyllium:** A water-soluble fiber; alleviates symptoms of constipation.

**Relative risk:** A comparison of the risks between two different groups.

**Saturated fats:** Cholesterol-raising fats, hard at room temperature, such as butter and lard.

**Serotonin:** A brain neurotransmitter that helps inhibit impulsive behavior. In some animal studies, low cholesterol levels have been linked with low levels of serotonin.

**Simvastatin (Zocor):** A cholesterol-lowering drug that interferes with the body's ability to make cholesterol in the liver; an HMG CoA reductase inhibitor.

**Stearic acid:** A component of the saturated fats found in red meat and chocolate, thought not to raise LDL cholesterol levels.

**Sterols:** The class of compounds to which cholesterol belongs.

**Trans fatty acids:** Compounds found in hydrogenated fats that have the same cholesterol-raising properties as saturated fats.

**Triglycerides:** Fatty compounds found in the blood that contain three chains of fat and one glycerol (alcohol) molecule.

**Vasospasms:** Contractions of the blood vessels.

**Very-low-density lipoprotein (VLDL):** A type of **lipoprotein** converted in the body to LDL cholesterol and triglycerides. Most forms of VLDL do not appear to play a role in the development of atherosclerosis.

**Xanthomas:** Fatty deposits of cholesterol in tendons and elsewhere, associated with inherited cholesterol disorders.

# SUGGESTED READING

Demrow, Heather S., et al. "Administration of Wine and
  Grape Juice Inhibits In Vivo Platelet Activity and
  Thrombosis in Stenosed Canine Coronary
  Arteries." *Circulation* 91 (February 15, 1995):
  1182-1188.

"Detection, Evaluation and Treatment of High Blood
  Cholesterol in Adults (Adult Treatment Panel II)."
  *Circulation* 89 March 1994): 1330-1445.

"Effect of Estrogen or Estrogen/Progestin Regimens on
  Heart Disease Risk Factors in Postmenopausal
  Women." Writing Group for the PEPI Trial.
  *Journal of the American Medical Association*
  273 (January 18, 1995): 199-207.

Glantz, Stanton A., M.D. "Passive Smoking and Heart
  Disease." *Journal of the American Medical
  Association* 273 (April 5, 1995): 1047-53.

Hulley, Stephen B., M.D., et al. "Should We Be Measuring Blood Cholesterol Levels in Young Adults?" *Journal of the American Medical Association* 269 (March 17, 1993): 1416-1427.

Jacobs, David, Ph.D., et al. "Report of the Conference on Low Blood Cholesterol: Mortality Associations." *Circulation* 86 (September 1992): 1046-1060.

Kaplan, Jay R., et al. "Demonstration of an Association Among Dietary Cholesterol, Central Serotonergic Activity and Social Behavior in Monkeys." *Psychosomatic Medicine* 56 (November/December 1994): 1-5.

Krumholz, Harlan M., et al. "Lack of Association Between Cholesterol and Coronary Heart Disease Mortality and Morbidity and All-Cause Mortality in Persons Older Than 70 Years." *Journal of the American Medical Association* 272 (November 2, 1994): 1335-1340.

Kwiterovich, Peter, M.D. *Beyond Cholesterol: The Johns Hopkins Complete Guide for Avoiding Heart Disease.* Baltimore, Md.: The Johns Hopkins University Press, 1989.

Lauer, Ronald M., M.D., and William R. Clarke, Ph.D. "Use of Cholesterol Measurements in Childhood for the Prediction of Adult Hypercholesterolemia." *Journal of the American Medical Association* 264 (December 19, 1990): 3034-3038.

Law, M.R., et al. "Assessing Possible Hazards of
Reducing Serum Cholesterol." *British Medical
Journal* 308 (February 5, 1994): 373-379.


Levine, Glenn N., M.D., et al. "Cholesterol Reduction
in Cardiovascular Disease." *The New England
Journal of Medicine* 332 (February 23, 1995):
512-519.


Moore, Thomas J. "The Cholesterol Myth." *Atlantic
Monthly* (September 1989): 37-70.


Muldoon, Matthew F., et al. "Effects of Acute Psychological Stress on Serum Lipid Levels, Hemocon-
centration and Blood Viscosity." *Archives of
Internal Medicine* 155 (March 27, 1995):
615-620.


Muldoon, Matthew F., et al. "Low or Lowered
Cholesterol and Risk of Death From Suicide and
Trauma." *Metabolism* 42, suppl. 1, (September
1993): 45-56.


National Cholesterol Education Program, "Report of
the Expert Panel on Blood Cholesterol Levels in
Children and Adolescents." NIH Publication
No. 91-2732.


Ornish, Dean, M.D. *Dr. Dean Ornish's Program for
Reversing Heart Disease.* New York: Random
House, 1990.

Scandinavian Simvastatin Survival Study Group.
"Randomized Trial of Cholesterol Lowering in
4444 Patients With Coronary Heart Disease:
The Scandinavian Simvastatin Survival Study
(4S)." *Lancet* 344 (November 19, 1994):
1383-1389.

Simon, Harvey B., M.D. *Conquering Heart Disease:
New Ways to Live Well Without Drugs or
Surgery.* Boston.: Little, Brown, 1994.

Vine, Donald L., M.D. "Doubts About Clinical Impact
of Cholesterol Reduction." Reply from William
Castelli, M.D. "The Folly of Questioning the
Benefits of Cholesterol Reduction." Dr. Vine
responds to Dr. Castelli. *American Family
Physician* 49 (February 15, 1994): 558-547.

# INDEX

## A

Absolute risk, defined, 37, 169

Age

  cholesterol and, 29-30

  effects, 22, 23, 26, 60, 80-82

Alcohol, effects, 86, 112-113

American Heart Association

  information, 167

  low-fat diets, 114-118

Angina, age and, 22

Angina pectoris, defined, 31, 169

Antioxidant, defined, 106, 169

Antioxidant vitamins, 17, 94, 120-121

Apolipoproteins, defined, 78, 169

Arteriography, defined, 92, 169

Arteriosclerosis, defined, 15, 169

Aspirin, effects, 112

At-home tests, 45-46

Atherosclerosis

  age and, 22

  cholesterol level and, 14

  defined, 14-15, 170

  diet and, 94

  studies on cholesterol and, 27-30, 32-34

Atromid-S. *See* Clofibrate (Atromid-S)

Attributable risk, defined, 38, 170

## B

Behavior problems, low cholesterol levels and, 87-88

Beta-carotene

  defined, 94, 170

  effects, 94

Bile acid binding resins

  in children, 76, 132

# THE SIGN OF FOUR

broadview editions
series editor: L.W. Conolly

Photograph of Arthur Conan Doyle (1890), taken by Herbert Rose
Barraud (1845-96). Published in *Men and Women of the Day 1893*
(London: Eglington & Co., 1893).

**Library and Archives Canada Cataloguing in Publication**

Doyle, Arthur Conan, Sir, 1859-1930
      The sign of four / Arthur Conan Doyle ; edited by Shafquat Towheed.

(Broadview editions)
Includes bibliographical references.
ISBN 978-1-55111-837-6

      I. Towheed, Shafquat, 1973-   II. Title.   III. Series: Broadview editions

PR4622.S54 2010              823'.8        C2010-902616-0

**Broadview Editions**
The Broadview Editions series represents the ever-changing canon of literature in English by bringing together texts long regarded as classics with valuable lesser-known works.

Advisory editor for this volume: Michel Pharand

Broadview Press is an independent, international publishing house, incorporated in 1985.

We welcome comments and suggestions regarding any aspect of our publications—please feel free to contact us at the addresses below or at                          ,
broadview@broadviewpress.com.

*North America*
Post Office Box 1243, Peterborough, Ontario, Canada K9J 7H5
2215 Kenmore Avenue, Buffalo, NY, USA 14207
Tel: (705) 743-8990; Fax: (705) 743-8353
email: customerservice@broadviewpress.com

*UK, Europe, Central Asia, Middle East, Africa, India, and Southeast Asia*
Eurospan Group, 3 Henrietta St., London WC2E 8LU, United Kingdom
Tel: 44 (0) 1767 604972; Fax: 44 (0) 1767 601640
email: eurospan@turpin-distribution.com

*Australia and New Zealand*
NewSouth Books
c/o TL Distribution, 15-23 Helles Ave., Moorebank, NSW, Australia 2170
Tel: (02) 8778 9999; Fax: (02) 8778 9944
email: orders@tldistribution.com.au

www.broadviewpress.com

This book is printed on paper containing 100% post-consumer fibre.

Typesetting and assembly: True to Type Inc., Claremont, Canada.

PRINTED IN CANADA

# THE SIGN OF FOUR

## Arthur Conan Doyle

*edited by Shafquat Towheed*

broadview editions

# Contents

# Acknowledgements

I am grateful to the British Library for granting permission to reproduce the cover image, from the Maurice Vidal Portman collection, Andamanese Islanders, Vol. VII, APAC Photographic Archive, photo 188/7(25). The staff in the Asia, Pacific, and Africa Collections (APAC) and the Rare Books and Music Room at the British Library were unfailingly helpful. I would like to thank my graduate students at the Open University and the Institute of English Studies (University of London) and undergraduate students at the Florida State University London Study Centre. Thanks also to Kirsten MacLeod for first suggesting I take this project to Broadview, and to the editorial, production, and promotion team at Broadview, especially Leonard Conolly, Julia Gaunce, Tara Lowes, Marjorie Mather, and Michel Pharand for bringing this edition to life.

# Introduction[1]

## The Creator of Holmes

Born in Edinburgh on 22 May 1859 as the third of the nine children of Charles Altamont Doyle, an alcoholic draughtsman and artist, and his bookish wife Mary Foley, Arthur Conan Doyle was an unlikely candidate to have created one of the most enduring literary characters of British fiction. The Doyles were of Irish Catholic stock and of uncertain economic means; owing largely to his father's hopeless alcoholism, the young Arthur Conan Doyle's childhood was both financially and economically precarious. Benefiting from the largesse of rich relatives, he attended Stonyhurst College (1870-75), Britain's top Jesuit school, before returning to Edinburgh (1876-81) to study medicine. Like so many other students at Scotland's finest university, the young Doyle wanted to do more than complete his professional training: he also wanted to be a writer. In 1879, his first short story, "The Mystery of the Sasassa Valley," was published in *Chambers's Edinburgh Journal*, earning him a paltry £3 3s; the same year saw his first non-fiction, "Gelseminum as a poison," appear in the *British Medical Journal*. Safely established in general practice in Portsmouth, Doyle continued to foster his nascent literary career: "The Captain of the Pole-Star," derived largely from an earlier trip to the Arctic as a ship's doctor, appeared in the pages of the highly respected *Cornhill Magazine* in January 1884.

At this point, and indeed, for the remainder of his literary career, Arthur Conan Doyle's ambition was to become a respected writer of historical fiction. However, events took an unexpected turn with the publication of his first Holmes and Watson story, "A Study in Scarlet," in the influential pages of *Beeton's Christmas Annual* for 1887. The story was a sales success, and Doyle found that he could earn ready money by turning his hand to detective fiction. When he was commissioned by Joseph M. Stoddard, the editor of the influential and popular American literary journal *Lippincott's Monthly Magazine*, to write a short novel, he found himself being dined (and sharing space on the page) with Oscar Wilde (Stoddard had also commissioned *The*

---

1  Readers should note that this Introduction gives away major plot points.

*Picture of Dorian Gray* for *Lippincott's*). Doyle received £100 for the serialisation, equivalent to a third of his annual income from practising medicine.

*The Sign of Four*, then, marks a potential bifurcation in Doyle's literary career. Was he ordained to become a celebrated writer of historical romances, along the lines of his fellow Scotsmen Walter Scott and Robert Louis Stevenson before him? Or was his fictional creation at 221b Baker Street destined to eclipse not just Doyle's ambition to be a historical novelist, but even its creator? At the time that Doyle wrote *The Sign of Four*, the outcome was far from obvious, and his own authorial investment in the Holmes and Watson stories was circumspect rather than enthusiastic. The runaway success of the next six Holmes and Watson stories in the pages of the *Strand Magazine* in 1891 changed all of that. Both critical and commercial success was now guaranteed, but more importantly, Doyle had unwittingly created a literary phenomenon that even to this day shows no signs of abating.

## The "Science" of Deduction

*The Sign of Four* remains one of the best known and most widely read of Arthur Conan Doyle's Holmes and Watson stories. Holmes's most famous dictum, that "when you have eliminated the impossible, whatever remains, *however improbable*, must be the truth," appears not once, but *twice* in *The Sign of Four*, the only Holmes story where it is repeated. In Holmes's re-introduction in front of the reading public, the narrative served to impress upon them the particular quirks of his personality and offered the clearest exposition of his working methods. Doyle deliberately named the first chapter of the novel "The Science of Deduction," encouraging readers who had already encountered Holmes to connect *The Sign of Four* with the title of the second chapter of *A Study in Scarlet* (1887). *The Sign of Four* was published to positive reviews, but initially had relatively modest sales; it was not until the appearance of the first of the Holmes and Watson stories, "A Scandal in Bohemia," in the newly launched *Strand Magazine* in July 1891, immediately after the passage of the first reciprocal Anglo-American copyright law, that the world's most famous consulting detective achieved significant sales. Following the success of the stories in the *Strand*, *The Sign of Four* was reprinted and reissued at an accelerated rate (see Note on the Text), reaching a wider and more diverse readership. However, despite the confidence with which Holmes is depicted and the case expounded, Holmes's future as a fictional device was far from certain in 1890; Doyle still considered himself a historical novelist and *not* a writer of detective fiction.

One of the most interesting features of Doyle's novel is the extent to which it draws upon contemporary debates in fields such as criminal anthropology, anthropometry (the classification of individuals and population groups through physical measurement), physiognomy, and psychology. Havelock Ellis's work on criminal anthropology, the core of which was presented at the Second Congress of Criminal Anthropology, held in Paris from 10 to 17 August 1889, argued that specific physical characteristics indicated an innate predisposition to crime. Ellis's premise was based on the principle of atavism (evolutionary regression) and drew heavily upon the work of the pioneering Italian criminal anthropologist Cesare Lombroso (1835-1909), especially his *L'uomo delinquente* (*Criminal Man*, 1876). In *The Criminal* (see Appendix A1), Ellis argued that characteristics like prognathism (the prominence of the lower jaw), hair and eye colour, skin complexion, hirsuteness, and the prevalence of wrinkles, could be used to identify a criminal type or an individual's preponderance to commit crime. "The lower jaw," Ellis argued, was "well developed in those guilty of crimes of violence," for "the criminal resembles the savage and the prehistoric man" (Ellis, 64). According to Ellis, criminals were hairier and darker than usual. "The abundance of hair seems to be correlated with the animal vigour which is often so noticeable among criminals," he observed, adding that it could be "explained by arrest of development or atavism" (Ellis, 72-4). Ellis noted that the faces of convicted criminals were more heavily lined with wrinkles than average, and examining the current Scotland Yard statistics for the hair colour of wanted criminals, concluded that a disproportionate number of dangerous criminals had dark hair, and very few were red-headed: "of 129 persons 'wanted' at Scotland Yard," Ellis observed, "45 have 'dark brown' hair, and of these 17 (*i.e.*, 37.7 per cent) are described as 'dangerous,' 'desperate,' 'expert,' or 'notorious,'" adding that "as exact evidence on the colour of the hair goes, it points chiefly to a relative deficiency of red-haired persons among criminals" (Ellis, 76-7). Athelney Jones, who in *The Sign of Four* arrests the red-haired Thaddeus Sholto for the murder of his twin brother Bartholomew, is clearly not up to date with either his own police statistics, or familiar with Ellis's work. Another character, Small, is repeatedly referred to as wild, hairy, fierce, brown faced, and even as a "brown, monkey-faced chap" (103), constantly drawing our attention to his regressive, depraved, and animalistic state.

Watson, as Lawrence Frank has observed, is "a man of Galtonesque, Lombrosian persuasion" (Frank, 177), but it also needs to be recognised that the arbiter of useful data, Holmes, does not dispute the claims made by Ellis or Lombroso: Holmes's monograph on the "influence of a trade upon the form of the hand" (53) mimics the discourse of determinist taxonomy. While Small represents one readily diagnosed physiological and psychological type (the recidivist criminal), Holmes himself represents another (the genius). Perhaps the most overt depiction of genius in the period, Holmes glosses many contemporary concerns about defining and explaining genius, seen in books such as Lombroso's *The Man of Genius* (1891) and Francis Galton's *Hereditary Genius* (1869). Lombroso considered genius to be a "special morbid condition" (Lombroso, v) akin to insanity, typified by a dependence on narcotics, a love of difficulty, extreme egotism, bipolarity, and a disposition to psychosis (see Appendix A2).

Holmes displays all these traits. At the beginning of the narrative, he injects himself with a seven percent solution of cocaine in a desperate attempt to stave off boredom, for he has a mind that "rebels at stagnation" (50). Once involved in Mary Morstan's case, Holmes displays both physiological and psychological bipolarity; he remains in a state of heightened manic excitement, and does not sleep for 82 hours. Once the case is solved, he confesses that he will descend to a depressive state and feel "as limp as a rag for a week" (156), which prompts Watson to observe that "what in another man I should call laziness alternate with your fits of splendid energy and vigour" (156). Watson frequently comments on Holmes's egotism and lack of empathy, and, alarmingly, suggests the congruity of criminality and genius: "I could not but think what a terrible criminal he would have made had he turned his energy and sagacity against the law, instead of exerting them in its defence" (85). The impetus towards typology and classification in *The Sign of Four*, something that both John McBratney and Yumna Siddiqi have commented upon, applies not just to the ostensibly criminal individual (Small), or to the allegedly degenerate race (Tonga), but also to the genius of Holmes. Doyle presents a cast of characters as a casebook of identifiable, indexed, and categorised types that would have been familiar to his readers.

*The Sign of Four* features the first occurrence in the Holmes stories of an activity that has since become the hallmark of detection the world over: the examination of fingerprints. Holmes notes the smudged thumbprint on Thaddeus Sholto's note to Mary Morstan (59), and carefully uses lens and tape measure to examine the evidence, including possible fingerprints, left in the dust of Sholto's attic (85). *The Sign of Four* fittingly shared its billing in the February 1890 number of *Lippincott's Monthly Magazine* with the statistician, eugenicist, and evolutionary scientist Francis Galton (1822-1911), whose article on anthropometry, "Why do we measure mankind?," was the third piece in that number. Galton had spoken on fingerprinting at the Royal Institution on 25 May 1888, so Doyle's discussion was particularly topical. Galton pioneered fingerprinting in Britain; *Finger Prints* (1892) concluded that no two fingerprints could be the same, and he introduced a system of classifying prints by examining individual loops, whorls, and arches.

The use of fingerprinting for classifying and identifying suspects originated not in Doyle's foggy London, but in rural Bengal. Working as a Bengal magistrate, Sir William James Herschel (1833-1917), the son of the famous astronomer, used hand and fingerprints to confirm contracts; on closer examination, he discovered that each individual's prints were unique, and could be identified against an archive. As magistrate of Hooghly, in 1877 he instituted the world's first corroborative fingerprint archive; developed by Galton, this was eventually adopted across India in 1897 and by Scotland Yard in 1901. The need to control a large and restless subject population meant that India was the site of pioneering work in British nineteenth-century criminology, from the use of photography to produce a criminal archive, to the use of anthropometry to identify potential criminals.

Lacking a corroborative archive, something he would have had in Bengal, Holmes's scrutiny of thumb and fingerprints amasses data without direct utility. It is telling that it is the imprint of Small's stump and the creosote-dipped mark of Tonga's undersized feet that provide the only trail in the narrative, and one that can only be followed by the keen nose of the bloodhound Toby. Dealing with the obtuse Athelney Jones, Holmes anticipates the potential of fingerprinting in securing convictions; the first successful criminal prosecution using fingerprint evidence, by Juan Vucetich in Buenos Aires, was in 1892.

Despite the valorising of the "science of deduction" in *The Sign of Four* (the word "deduction" occurs seven times), both the criminal processes and the conclusions that they support owe more to late nineteenth-century pseudo-science than anything else. *The Sign of Four* is valuable not because of the plausibility of the "science of deduction" in the resolution of the plot—the book is riddled with gaps, implausible information, and errors of logic, and Holmes's assumptions are sometimes breathtakingly sweeping—but because of the unerring accuracy with which it picks up and productively uses ideas from contemporary science and pseudo-science, such as anthropometry, criminal classification through physiognomy, and ongoing debates over the nature of human genius.

## Reading and the Suspension of Disbelief

Despite presenting us with enough information to construct an internal chronology, Doyle's novel poses inconsistencies that are impossible to reconcile. The narrative ostensibly opens on the afternoon of 7 July 1888 **(day 1)**, with Mary Morstan's arrival at 221b Baker Street with a letter, postmarked earlier that day, inviting her to a meeting outside the Lyceum Theatre at 7 pm the same evening. After travelling to Thaddeus Sholto's house in Brixton, Holmes and Watson visit Bartholomew Sholto's house in Upper Norwood, where they arrive at nearly 11 pm (Chapter 5). Soon after, they discover Bartholomew Sholto's "stiff and cold" body, which is already in an advanced state of *rigor mortis*, caused not just by the time elapsed since death, but by a powerful alkaloid poison. Sholto's body is cold, confirming that the murder took place over 24 hours before **(day 0**, after 10 pm the night before, when Thaddeus left his brother's house), but the creosote marks left by Tonga are still fresh enough for Watson to smell, and for Toby to follow.

Watson escorts Mary Morstan back to Mrs. Cecil Forrester's at Camberwell at 2 am the next day **(day 2)**, collects Toby, and brings him to Pondicherry Lodge just after 3 am. By the time Holmes and Watson set off with Toby on their seemingly fruitless six-mile trudge (Chapter 7), the "east had been gradually whitening" and they can already see "some distance in the cold grey light." Holmes has a moment of epiphany seeing the sunrise, poetically comparing clouds to flamingos, and by the time they reach Mordecai Smith's wharf (Chapter 8), they see Mrs. Smith attempting to wash her son, clearly in broad daylight (Mrs. Smith tells us that Small woke her husband "yesternight" at 3 am). Holmes and Watson return to 221b Baker Street between 8 and

9 am, having stopped at a post office en route; they read the first news report of Bartholomew Sholto's murder in the *Standard* over breakfast. After the visit of the "irregulars," Holmes lulls Watson to sleep with his violin improvisation; he wakes late in the afternoon, visits Mary Morstan at Camberwell, and returns again after dark to find Holmes pacing frantically (Chapter 9). Holmes continues to pace through the night, emerging the next day **(day 3)** for breakfast with a "little fleck of feverish colour." Nothing further takes place that day; Watson visits Mary at Camberwell, and Holmes throws himself into a chemical experiment, which continues until the "small hours of the morning."

Watson is woken at dawn the next day **(day 4)** by Holmes dressed as a sailor, who goes off to search for the *Aurora*, leaving Watson to answer messages. At 3 pm Athelney Jones arrives to request Holmes's help, and comments that "it is very hot for the time of year." Holmes returns heavily disguised as a sailor soon after, and invites Jones to dine with them "in half an hour." It clearly takes longer than that to prepare grouse and oysters for dinner, even for Sherlock Holmes, and by the time they finish their meal with a glass of port, it is 6.30 pm; they reach the Westminster wharf just after 7 pm (Chapter 10). It is twilight by the time they get to the Tower of London, and the chase takes place on a clear but dark night after 8 pm (when Small collects the *Aurora* from Mordecai Smith). Holmes and Watson shoot Tonga and capture Small, who offers a full, colourful, and highly contradictory confession the same evening. The action of the narrative, therefore, ostensibly takes place over 96 hours in July 1888, from the murder of Bartholomew Sholto after 10 pm on 6 July, to Small's confession late on the evening of 10 July; Holmes is on a cocaine-induced, sleep-deprived manic high for at least 82 of these hours.

So far then, so relatively straightforward, but the first complication is offered by the text itself. While Mary Morstan's note is dated "7 July," the narrator tells us in no uncertain terms that the journey to the Lyceum later that day was "a September evening" with the street lamps on the Strand already lit—clearly not something that would have happened in London in July, when it doesn't get dark until well after 9 pm. Doyle's poetically liberal use of moonlight and fog adds to the conflicting evidence over the *when* of the story; the swirling yellow fog is inconsistent and implausible for July (but not September), but the much more problematic references to moonlight (always strong) occur no fewer than seven times in the novel. If the action took place between 6 and 10 July 1888, the moon would be waning in its third quarter (6 July) before the new moon on 13 July; there would be some moonlight, but not much (and certainly less than the "half a moon" noted on the first trip to Pondicherry Lodge). Even more disturbingly, if the narrative is set between 6 and 10 September 1888, the moon would be waxing in its first quarter, with the new moon on 6 September: in other words, a thin crescent moon, which certainly wouldn't offer the bright moonlight constantly referred to on the night at Pondicherry Lodge. The choice is between a partially moonlit but fog-free July, or a nearly moonless but foggy September, and the evidence offered by Doyle's story resolutely refuses to let us make a definite choice.

Further complications arise if we try to determine the days of the week. If we accept 6 to 10 July 1888 for the events of the story, then Sholto's murder takes place on the evening of Friday 6 July, Mary Morstan arrives with information about the case on Saturday afternoon (7 July), the news of Sholto's murder is reported in the *Standard* on Sunday morning (8 July), and Holmes's advert in the paper and the river chase takes place on Tuesday (10 July). If we accept 6 to 10 September 1888, Sholto's murder takes place on Thursday evening (6 September), Mary Morstan arrives on Friday afternoon (7 September), the news of Sholto's murder is reported in the Saturday morning paper (8 September), and Holmes's advert and the river chase take place on Monday (10 September). Neither set of dates can be reconciled with the details of the disappearance of the *Aurora* in Holmes's advert in the *Standard*, which clearly states that the boat "left Smith's Wharf at or about three o'clock last Tuesday morning."

More conflicting evidence is offered by Holmes's choice of dinner on day 4 (impossible in July—the grouse season starts on 12 August, and before widespread refrigeration oysters were only eaten from September to April), Thaddeus Sholto's choice of coat (an Astrakhan frock coat and hat with lappets— plausible for an invalid in September, but ludicrous in July), Holmes's interpretation of *when* Mordecai Smith left with Small on the *Aurora* (just after 3 am would be pitch dark in September, but nearly light, which is what Holmes suggests, in July), Mrs. Forrester's decision to go out late in the evening on day 4 (the London social season ran from May to July—why would a respectable lady be out *without* her companion in September?), Athelney Jones's comment about the weather ("very hot for the time of year"— plausible for September but not worthy of comment in July), and the fact that twilight at Tower Bridge on day 4 is around 8 pm, which is accurate for September, but impossible for July. Even though the action of the story purports to be linear, continuous, and diachronic, the evidence it offers poses formidable problems of analysis.

The broader chronological context offers just as many insoluble problems, especially in the narratives of Major Sholto and Jonathan Small, but has attracted rather less scrutiny. Major Sholto's deathbed confession, as related to us by Thaddeus (the *only* surviving member of the entire family at the start of the narrative), suggests the following sequence of events, based on the assumption that the story is set in 1888. The Sholto twins, Bartholomew and Thaddeus, are born in 1858, for Watson tells us that Thaddeus has "just turned his thirtieth year," although it is not clear whether they are born in England or in India immediately after the "Mutiny." After serving in the Andamans, Sholto retires to Pondicherry Lodge with a large retinue of Indian servants in 1877, immediately after the end of his Army life, and bringing with him both the Agra treasure and a considerable inheritance from an uncle. On the evening of 3 December 1878, Morstan arrives in Pondicherry Lodge, the two men argue, and Morstan dies; neither of his 20-year old sons seems to know about this, so presumably they were both away at university at the time (Bartholomew is a good mathematician and chemist). Sholto's butler Lal Chowdar helps him dispose of Captain Morstan's body, which is never found; astonishingly, neither Holmes nor Mary Morstan show any interest whatever in locating it. Lal Chowdar, who dies sometime between Morstan's death (3 December 1878) and Sholto's (28 April 1882), takes this secret with him to the grave.

Sholto receives a distressing (and still unexplained) letter from India early in 1882, and by April of that year, is dying from an enlarged spleen. On 28 April 1882 he confesses his theft of Morstan's treasure to his sons, only to be frightened—literally to death—by the appearance of Jonathan Small at the window. It takes the brothers over six years (until Bartholomew's announcement the day before the start of the story, and possibly on the very day he was murdered) to find the treasure hidden in the false roof in Pondicherry Lodge. Sholto's account is seemingly coherent, except for that all too Gothic literary device, the distressing letter. Who sends this short letter from India in a "scrawling hand" early in 1882, and why is Sholto scared half to death by it? Why did Morstan, according to Sholto, not tell anyone other than

Sholto about his weak heart, when there was a resident surgeon at the penal settlement?

Small's confession is, if anything, even more unreliable than Sholto's. Born around 1837, Small enlists in the Army as an 18-year old to escape a "mess over a girl" (presumably a case of rape, or pregnancy, or perhaps, both) and is sent to India. Improbably, he loses his right leg to a crocodile while swimming in the Ganges, and after five months in hospital is discharged from the army, a "useless cripple," short of his twentieth birthday. No sooner does he become an overseer in the newly arrived Abel White's indigo plantation in Mathura than the Indian "Mutiny" breaks out (June 1857). This part of Small's account, cobbled together by Doyle from a variety of published sources (see Appendix B), is chronologically relatively accurate; the rest is riddled with inconsistencies. If Small serves twenty years (c. 1858-78) in the penal settlement before escaping, why does he not make common cause with Captain Morstan, who returns to England in December 1878, in confronting Sholto? Morstan and Small continued to communicate about the Agra treasure in the Andaman settlement after Sholto's departure, for Small admits that Morstan had shown him the "list of passengers" with Sholto's name, confirming that he had left for England.

Given the conflicting (and, perhaps, unreliable) accounts of Small and Sholto, a number of major issues remain impossible to resolve. Perhaps the biggest is the infamous issue of Small's face at the window on the night of Sholto's death: 28 April 1882. Small claims that after escaping from the Andaman Islands, he finally arrived in England only "three or four years ago" (i.e., 1884 or 1885), which would make this encounter impossible. If Small is so unreliable about one of the most important events in the narrative, how can we trust his evidence for anything else? Why, for example, is Small sentenced to death (later commuted to transportation for life) when his collaborators in the killing of Achmet are not? Small tells us that he has spent twenty years as a prisoner in the Andamans before escaping, and also that because of his good behaviour he owned his own hut and became a trusted petty officer, serving as an assistant to the surgeon. Under the "ticket of leave" system (see Appendix D5), even prisoners transported for life to the Andaman Islands were eligible for manumission after they had served twenty-one years, so why did Small bother to escape when he was almost on the point of being released? If Small did indeed have a "confederate" (Lal Rao) inside Pondicherry Lodge, why did he find it so difficult to get access to Sholto? And if the body of Achmet was discovered within a few days of the murder, why were the jewels not found, when the conspirators had buried the box in the "same hall" as the corpse? And, most obviously, why does Small wander the world, earning a living by parading Tonga, instead of heading straight to Pondicherry Lodge to reclaim his treasure and get his revenge against Sholto?

Finally, even Mary Morstan's seemingly innocuous chronology offers inexplicable gaps. She is born in India in 1861, and sent "home" to a "boarding establishment" in Edinburgh when still a child (she would have to have been at least five, and more likely seven, to have been accepted as a full-time boarder). Her father disappears in December 1878, when she is 17 years old; she enters the service of Mrs. Cecil Forrester as a governess (or, possibly, as a paid companion) in April 1882, presumably soon after achieving her majority (21) without an inheritance. What she does without a father, an inheritance, or an employer for the three and a half years between these dates is a mystery left unexplained by the narrative; neither Holmes nor her putative fiancé Watson seem to bother to find out. When Watson proposes marriage to her at the end of *The Sign of Four*, she is 27 years old, already old for a first-time bride in Victorian England.

The inconsistencies of *The Sign of Four* are legion, and the novel requires from the reader more than the usual level of suspension of critical disbelief. Even Holmes is complicit in the make-believe of deduction. Most famously, both the genius Holmes and his idiot double Athelney Jones fail to establish the identity of the "confederate" inside Pondicherry Lodge, or provide a plausible reason for his behaviour. Jones's arrest of Lal Rao is, as Holmes gleefully reminds us, accidental: "Jones actually has the undivided honour of having caught one fish in his great haul" (156). While Jones's omission is forgivable, Holmes's own inability to match Lal Rao with the "confederate" who, he surmises, must have been an accomplice to the murder of Bartholomew Sholto until the end of the case, is not. There are at the very least three Indian servants mentioned in *The Sign of Four*: Major Sholto's late butler and accomplice, the trusted Lal Chowdar; Thaddeus Sholto's nameless *khitmutgar*, who formerly worked for Major Sholto; and Bartholomew Sholto's butler Lal Rao, who proves to be instrumental in the murder. Entirely living up to nineteenth-century prejudices about the untrustworthiness of Indian witnesses, Holmes makes no attempt to question either of the two living butlers, despite the fact that Lal Rao may well have been the only witness to the murder of Bartholomew Sholto. Holmes's line of questioning here is clearly not determined by the "science of deduction," but by the assumptions of late nineteenth-century criminology, shaped by complicity in enforcing British authority on an unruly and potentially insurgent subject population.

Leslie Klinger[1] offers a summary of the various conflicting start dates already suggested for the opening of the story (Holmes scholars have offered at least seven, and none are entirely satisfactory), while William Baring-Gould[2] advances perhaps the most plausible argument (and certainly one of the most entertaining ones) for the action taking place between 18 and 21 September 1888, but says nothing about the many chronological inconsistencies in the narratives of John Sholto or Jonathan Small. In the final analysis it must be conceded that despite many references to actual historical events, and a plethora of seemingly concrete pieces of internal evidence, *The Sign of Four*, like all works of imaginative fiction, demands from the reader a considerable suspension of disbelief.

## "Mutiny"

Sixty years after Indian independence, and a hundred and fifty after the largest military and civilian struggle against an occupying Imperial power in the nineteenth century, the Indian "Mutiny" casts its shadow over the relationship between Britain and India. It was certainly the most important encounter, however asymmetrical, between the two nations in that century. For both countries, the events of the "Mutiny" shaped and re-shaped ideas about national, racial, religious, political, and gender identity. Nowhere were these ideological debates more strongly depicted and contested than in the fiction of the half-century that followed. *The Sign of Four*, I would argue, is intrinsically not just a work of detective fiction, but a response to (and reworking of) the profound impact of the events of 1857-58 on the collective British imaginative consciousness.

A note here about terminology: I have preferred to use the term Indian "Mutiny" to indicate the still evolving and histori-

---

1 Leslie S. Klinger, ed., "The Dating of *The Sign of Four*," *The New Annotated Sherlock Holmes* (New York: W.W. Norton, 2006), 380-81.

2 William Baring-Gould, *The Chronological Holmes* (privately printed, 1955), 8-13.

cally contested nature of the interpretation and historiography of the series of events that took place in 1857-58. Even the most partisan of British historical accounts had by the late nineteenth century stopped referring to it as the "Sepoy Mutiny," for this failed to acknowledge the prevalence of discontent (and misrule) in the East India Company's administration. Imperial historians were forced to concede that this was more than a mutiny of Indian soldiers rebelling over the use of animal fat for greasing cartridges; note for example, the development of the title of Kaye and Malleson's multivolume tome from *A History of the Sepoy War* (1864-65) to *A History of the Indian Mutiny* (1888-89). Similarly, the current Indian nationalist vogue (one contested by the majority of Indian historians) for the expression "The First War of Independence," suggesting coordination, premeditation, and unanimity of purpose, is potentially profoundly misleading. The vast majority of combatants on both sides of the conflict during 1857-58 were Indians, as were the overwhelming majority of its victims, caught up in combat and the reprisals that took place afterwards. "Uprising," a term currently gaining favour among historians and literary scholars alike, is perhaps the most accurate and appropriate, but I have not used it because it lacks the metonymic immediacy and charge that the word "mutiny" so clearly carried at the time in both the British and Indian imaginations.

Thirty years (incidentally, roughly the traditional period of antedating narrative that separated the genres of the contemporary and the historical novel) after the events of 1857-58 saw the publication of one of its most important official histories, John William Kaye and George Bruce Malleson's six-volume cabinet edition of *The History of the Indian Mutiny of 1857-8* (see Appendix B6). Issued at the rate of one volume every two months between 1 October 1888 and 1 October 1889, the publication of Kaye and Malleson's *History* immediately precedes *The Sign of Four*, and its target readership would have been exactly the professional middle class (doctors, army officers, idle gentlemen, etc.) that populate the London of Doyle's novel. Kaye and Malleson were far from alone in attempting to capture the "Mutiny" for posterity, for the pace and range of writings it spawned in a variety of genres accelerated considerably in the last decade of the nineteenth century. As Gautam Chakravarty observes, "the most productive period in the history of the Mutiny novel was the 1890s," and these books tell us "a good deal about the British self-image in India, as they are among the indices of a high imperial culture" (Chakravarty, 6). Mutiny novels published in the 1890s include G.A. Henty's *Rujjub the Juggler* (1893) and *In Times of Peril: A Tale of India* (1899), H.C. Irwin's *A Man of Honour* (1896), A.F.P. Harcourt's *Jenetha's Venture* (1899), Flora Annie Steel's *On the Face of the Waters* (1896), and Lucy Taylor's *Sahib and Sepoy* (1897). These novels were largely chauvinist, often overtly populist, and sometimes they articulated "conspicuous demonstrations of racial superiority" (Chakravarty, 6).

*The Sign of Four* does not share the overt chauvinism of Henty's novels or the romance of Steel's. It does not endorse any particular political programme, nor does it function as a juvenile primer for empire-building; and yet, as well as being a detective novel, it is also a mutiny novel, for it consciously foregrounds the events in Agra in the summer of 1857 by making it central to the explication of the plot. "The Strange Story of Jonathan Small" clearly mimics both historical and fictional accounts of the "Mutiny," and Small is himself aware of the knowledge and expectations of his audience. "Of course you know all about it, gentlemen—a deal more than I do, very like, since reading is not in my line," he tells Holmes, Watson, and Athelney Jones, before declaring the alleged veracity of his testimony: "I only know what I saw with my own eyes" (135). Born in 1859, and never having visited the proverbial jewel in Britain's imperial crown, Doyle's own knowledge of the Indian "Mutiny" was culled entirely from secondary sources. However, some of his readers would have experienced the events first hand, or have received vivid accounts from their friends and relatives, such as William Muir's *Agra in the Mutiny* (Appendix B1), while others would have been eager readers of the wave of new publications in a variety of genres. Doyle's choice of topic was not merely timely or opportunistic, although it was clearly both of those, but rather another indication of the pervasive fascination with the Indian "Mutiny" at the apogee of empire. By the time Doyle came to write Jonathan Small's narrative, the Indian "Mutiny," and the mythologising of its events and participants (such as the reification of Henry Havelock as a Christian martyr; see Appendixes B3 and B4) already loomed large in the imaginative consciousness of British readers. In this light, it is worth interpreting Doyle's fictional depiction of Agra during the summer of 1857 by comparing it with some of the published historical sources and memoirs.

Shoehorned into the end of the "case" as a confessional captive narrative, "The Strange Story of Jonathan Small" is a criminal autobiography, drawing upon both British (Newgate novels) and Indian (Thuggee confessions)[1] established genres in its telling. Small's account is heavily dependent on Doyle's reading of historical sources about the "Mutiny"; both its derivativeness and its creative infidelities are worthy of comment. Narratives of colonial settlement are trajectories of either overweening ambition, or desperate escape; in the case of Jonathan Small, a poor white from Worcestershire, it is the latter. His Indian life begins at the age of 18, when, after getting into a serious "mess over a girl" which he "could only get out of" by "taking the Queen's shilling" (134), he arrives in India as a subaltern, the year before the outbreak of the "Mutiny." No sooner does he establish himself as a soldier, than he improbably loses a leg to a crocodile while swimming in the Ganges. Invalided out of the army, he is employed by a newly arrived indigo-planter, the symbolically named Abel White (who proves to be less able than white). Small becomes the overseer of White's plantation, located in "a place called Muttra, near the border of the Northwest Provinces" (136), and when not whipping the plantation workers into shape, he apparently engages in the kind of heart-warming social solidarity beloved by proponents of empire: "white folk out there feel their hearts warm to each other as they never do here at home" (135). Small's sense of social solidarity proves to be both cursory and misplaced, for he offers White no assistance once the "Mutiny" actually breaks out; "I could do my employer no good, but would only throw my own life away if I meddled in the matter" (136), he declares.

Intriguingly, much of Small's account of life in pre-1857 India is historically accurate; this is something that no commentator on *The Sign of Four* has noticed. Indigo dominated commercial activity in the East India Company's territory, and forced cultivation was the usual method; 26 of the 36 pages listing "manufactories" in *The New Calcutta Directory of the town of Calcutta, Bengal, the North-West Provinces* ... (Calcutta: F. Carberry for the Military Orphan Press, 1857) were devoted to listing indigo plantations and factories. The Indigo Planters' Association was established in Calcutta in 1854 to represent the burgeoning interests of planters, and indigo was rapidly becoming the largest agricultural contributor to tax revenue in the East India Company's territo-

---

1 Newgate (or Old Bailey) novels sensationalised the lives of criminals (whose biographies were published in the *Newgate Calendar*). Thuggees (from Hindi "thief" and Sanskrit "deceiver") were assassins notorious for their silent, efficient murders; their name is the origin of "thug."

ries (its rival was opium grown for export to China). Muttra, correctly Mathura, did indeed have an indigo plantation, the Oomerghur Concern, which was owned by the firm of Mackillop, Stewart and Co., and managed by a resident British manager, Thomas Churcher; the plantation survived the "Mutiny," albeit with a new resident manager (it is unclear whether Churcher shared Abel White's grisly fate). The forced cultivation of indigo resulted in widespread hunger, and was one of the many contributing resentments that caused the "Mutiny" of 1857 and a succession of peasant uprisings that followed. This grievance would find its voice in one of the most famous works of Indian literary resistance to British rule, Dinabandhu Mitra's play *Nil Darpan; or, the Indigo Planting Mirror* (1861). It was reprinted in London and Edinburgh in 1862; Small's cameo role as a soldier and indigo-plantation overseer rapidly sketches out the economic and political conditions leading to the "Mutiny," something that many of Doyle's readers would have noted.

Witnessing the destruction of the plantation by the "hundreds of black fiends ... dancing and howling round the burning house" (136), Small flees to Agra, where he joins his fellow "helpless fugitives" (136) in sheltering inside the Fort. Small joins the civilian "volunteer corps of clerks and merchants" and marches out to confront the rebel soldiers at "Shahgunge early in July" (137). While on guard duty in command of two newly recruited Sikh soldiers, Small is overpowered and, in a sacrilegious parody of the Sikh *Amrit Sanskar* ceremony that also clearly alludes to thugee initiation, he is forced to swear an oath that soon makes him an accomplice in the murder and robbery of the "unhappy merchant Achmet" (143). Small participates in Achmet's killing, and the body is buried in a "natural grave" inside a "great empty hall" (145) within Agra Fort; the treasure is secreted in a brick wall in the same room. The "Mutiny" ends, and the period of lawlessness that flourished during it is symbolically closed by the discovery of Achmet's body and the arrest of Small and his three Indian accomplices—Dost Akbar, Mahomet Singh, and Abdullah Khan—all of whom are at least nominally in the service of the British. The gang of four are tried, convicted, and sentenced, the three Indians to life imprisonment, and Small to death, later commuted to life. After short periods in Agra prison, and Madras, all four are transported for life to the penal colony at Port Blair in the Andaman Islands; the treasure remains securely hidden in Agra Fort, until it is purloined by the devious Major Sholto in 1877.

It is the conditions inside Agra Fort and the events taking place outside during the summer of 1857 that Doyle sketches most thinly, despite, or perhaps *because* of the wealth of extant historical documents (letters, memoirs, diaries, and histories) available to the reading public. We know (and many of Doyle's readers would have known) that the retreat of civilians and soldiers into Agra Fort took place rapidly on 1 July 1857, and that the city was recaptured by Colonel Greathed on 10 October but was not entirely under British control until the beginning of December; but unlike the unfortunate population holed up inside the residency in Lucknow, Agra Fort was *not* subject to a siege (see Muir, Appendixes B1 and B2 and Coopland, Appendix B5). The first problem the British administration inside the Fort faced was what to do about Agra prison, the largest penal settlement in the area, and already a British model prison for the whole of the territory under their control. Right at the outbreak of the "Mutiny," the Indian guards had deserted their posts; as the *de facto* administrator for Agra, Sir William Muir noted that their "position in Agra was in some degree complicated by having to guard our monster Jail by European troops, for the Jail nujeebs (armed guard) had gone off in a body towards the end of the month" (Muir, *Agra in the Mutiny*, 22). Muir's solution was creative, and would serve as a blueprint for future British policy; "the only resource left," he declared, was to "make over the custody of the jail to the Sikh prisoners, who were to be released and armed for the purpose" (Muir, *Agra in the Mutiny*, 23).

Ever since their defeat at the battle of Chillian Wallah during the Second Anglo-Sikh War in 1849, Sikhs had been recruited in increasingly large numbers to serve in the British administration as soldiers, guards, and police; this process increased markedly after the demonstrated loyalty of newly recruited Sikh soldiers during the events of the "Mutiny." As early as 1840, Henry Havelock, then embroiled in the First Anglo-Afghan War, pointedly noted that "there can be no medium, therefore, in the character of our relations with the Sikhs; they must either be established on a footing of the closest intimacy, or change at once into avowed hostility" (Havelock, *Narrative of the War in Affghanistan*, II, 233). Havelock's vow of instantly sworn friendship or implacable hostility is exactly the existential reality that Small encounters in his compact with the leader of the "Sikhs," Abdullah Khan: "Either you are heart and soul with us on your oath on the cross of the Christians, or your body this night shall be thrown into the ditch, and we shall pass over to our brothers in the rebel army" (140), he informs the hapless Small. Small's choice, as Havelock's had been, is one of expediency and complicity rather than principle; he agrees to join his Indian accomplices in murdering Achmet and stealing the Rajah's treasure. In a moment of astonishing candour, Abdullah Khan, the "taller and fiercer of the pair," tells Small that "we only ask you to do that which your countrymen come to this land for. We ask you to be rich" (140).

Small's account that they "went out to meet the rebels in Shahgunge early in July" (137) is based on a real enough event, which took place on 5 July 1857, but much of what happened in Agra in early July was rather less heroic and altogether more typical of the events of the "Mutiny," and remains unglossed by Small. In an attempt to re-impose their authority on the civilian population of the rebellious city, the British engaged in a punitive show of force, as William Muir vividly recounts: "On Wednesday (8th) a demonstration was made by marching a column through the city, and (I regret to say) by plundering the shop of a large Mahometan merchant in the military bazaar" (*Agra in the Mutiny*, 34). In fact, British looting was as widespread in Agra as Indian looting. As Ruth Coopland noted, "the officers sometimes made parties to go into the city and loot; but so great was the devastation, that they never brought us back anything, except a few cups and saucers and a coffee-pot" (Coopland, 186). British officers participated in looting Agra, while their newly recruited Sikh soldiers felt they had the right to take what they could as a reward for their loyalty, as Coopland observed: "They said it was very hard that they were not allowed to 'loot' Agra, as it was such a rich city" (Coopland, 231-32). Sometimes, the British and their allies were successful, as Coopland's apocryphal account of a "Seik finding some jewels of great value" (Coopland, 232) suggests, but more often than not, like Captain Garstone's systematic looting of Delhi Palace after the crushing of the "Mutiny" there, they found that the "splendid casket of ebony and mother of pearl, full of secret drawers" (Coopland, 258) had already been emptied. Watson's crestfallen discovery near the end of *The Sign of Four* that the Benares metal box was "absolutely and completely empty" (131) re-enacts a common enough British experience in India in the nineteenth century.

In fact, looting was so widespread during and immediately after the "Mutiny" that the possibility that the Agra treasure remained undiscovered inside the Fort for decades is highly implausible. The horrifying spectacle of the complete breakdown of law and order during the "Mutiny" was not only the rhetorically uncomplicated story of Indian rebellion and disorder, but the much more disturbing unravelling of British discipline, solidarity, and equanimity, a far cry from the officially sanctioned ideology of a just war conducted by a Christian army determined to restore order (see Grant, Appendix B3, and Williams, B4). Guarding Agra Fort from the threat of attack by the mutineers, both Small and his curiously heterogeneous "Sikh" confederates (or are they thuggee deceivers?) present a convincing mimicry of defenders of British order, while helping themselves at any opportunity to material gain. "I should like to know how many fellows in my shoes," Small confidently asks, "would have refused a share of this loot when they knew that they would have their throats cut for their pains" (144). Small's account is refreshingly short of hypocrisy; his justification for his actions compares favourably to Ruth Coopland's suggested collective punishment of the inhabitants of Delhi (see Appendix B5) for their participation in the "Mutiny."

Unlike the prolonged encampment inside Agra Fort, the "Mutiny" was quashed in Mathura relatively quickly. William Muir, the head of intelligence in Agra, and perhaps the best informed witness of events at the time, noted that the "Mutiny" in Mathura lasted from 30 May to 5 June 1857, when the town was rapidly reoccupied and order restored. "The ease with which a Magistrate and Collector with a handful of men recovers his authority *after the Sepoys have gone*," Muir wrote, "shows the nature of the rising as a military one, and the source of our difficulty" (Appendix B2). In this light, Small's prolonged stay in Agra and reluctance to return to Mathura and take command of Abel White's plantation after the re-establishment of British rule smacks of both desertion and opportunism. Could it be that Small, like the newly emancipated Sikh prisoners turned guards, knew there were riches to be looted in Agra, and he was therefore loathe to leave the city?

While Small's description of Agra Fort—"a very queer place— the queerest that ever I was in" (138)—is relatively truthful in terms of its size, location, and complexity, he gives us little information about what life was like inside the complex during the summer of 1857. A range of accounts gives us an accurate picture that Small's narrative cannot. Small declares that the "modern part" of the Fort "took all our garrison, women, children, stores, and everything else, with plenty of room over" (138); in reality, Agra Fort was far fuller than he suggests, and most of the occupants were not British. Ruth Coopland lists the results of the official census on 26 July 1857, which "amounted to no less than 5845; of which 1989 were Europeans, consisting of 1065 men and 924 women and children: the whole of the rest being natives and half-castes" (Coopland, 172). Indians and Anglo-Indians outnumbered the British inside the Fort by two to one, and this resulted in periodic outbursts of paranoia, mistrust, and recrimination. Coopland's memoir captures the embattled (and increasingly embittered) mood of the British towards Indians inside Agra Fort: "One baker was really hanged for planning a scheme for poisoning all the bread; and it was feared they might poison the wells" (Coopland, 188).

British resentment was exacerbated by the climate. The "Mutiny" broke out in May, the hottest month of the year, and reached its crescendo before the onset of that summer's monsoons. By all accounts, 1857 was hotter—and the monsoon later—than normal, but what made conditions even more unbearable for the British was the fact that they had no servants to fan them. As William Muir noted, "it was during the worst hot months of the year that we were thus incarcerated ... in the lack of servants, we had not even bearers to pull the Punkahs" (*Agra in the Mutiny*, 35). For Ruth Coopland, the insubordination of Indian servants inside Agra Fort took on the menacing spectre of insurgency. "The manners of the servants were most insolent and contemptuous," she observed; "the 'budmashes' [scoundrels] used to sing scurrilous songs under the walls, and draw pictures on them of the 'Feringhis' being blown up, with their legs and heads flying into the air; they also stuck up placards, saying on such a day we should all be massacred or poisoned" (Coopland, 188).

Small's narrative is silent about the human conditions inside Agra Fort that summer, but perhaps the greatest aporia in the novel is Doyle's cursory treatment of the immediate aftermath of the "Mutiny," which is glossed in a couple of sentences, with the assumption that his listeners are better informed of what actually happened than he is himself. "There's no use my telling you gentlemen what came of the Indian mutiny," Small tells his interrogators, for "a flying column under Colonel Greathed came round to Agra and cleared the Pandies away from it" (146). In fact, the British recapture of rebel towns was often accompanied by staged acts of retributive justice, together with calculated insults designed to outrage Indian religious sentiments. Ruth Coopland vividly describes the firing of captured mutineers from British guns outside Agra Fort: "One gun was overcharged, and the poor wretch was literally blown into atoms, the lookers on being covered with blood and fragments of flesh: the head of one poor wretch fell upon a bystander and hurt him" (Coopland, 233). James Grant's heavily evangelical biographical memoir of Henry Havelock offers a typical account of collective punishment, this time by General Neill in Kanpur. Neill made captured mutineers clean blood with their own hands before having them hanged and buried in a ditch; this was calculated to offend their religious (Hindu Brahmin) beliefs. Coopland gleefully narrates the Agra victory tiffin (washed down with alcoholic milk punch), designed to offend Muslim sensibilities, that took place "in one of the mosques of the Taj," where "all the ladies, children, officers, and soldiers were gathered; and here and there might be seen a native, looking green with rage at their sacred building being thus desecrated" (Coopland, 245). The collective punishment of Indians in the wake of the "Mutiny," as in the case of the forced removal of the entire civilian population of Delhi, was sometimes systematic, and usually recorded in the memoirs and correspondence of individual army officers, rather than in the official histories. Nevertheless we should not assume that Doyle's readers would have been entirely unaware of the nature of British retribution after the crushing of the "Mutiny." Small's assumption of the knowledge of these events in his interrogators (Holmes, Watson, and Athelney Jones) is entirely plausible.

Whether we interpret Doyle's silence about these acts of retribution in *The Sign of Four* as a sign of complicity or of censure remains a matter for debate. There is no extrinsic or intrinsic textual evidence in the novel to allow us to determine Doyle's ideological position, or whether he even occupied one, but the absence of any reference to the retributive aftermath of the "Mutiny" suggests the inability of this novel of detection to either contain or describe the events that took place. Doyle's own personal politics were firmly in favour of the union and empire; he stood twice (both times unsuccessfully) as a unionist candidate for Parliament, and was one of the leading advocates of both the Boer War and World War I (his political views are not evident in *The Sign of Four*). What Doyle's tale of colonial greed and retribution does suggest is that the heroic narratives of national and cultural identity woven around the events of the Indian "Mutiny" were open to interrogation. Even historians as sagacious as Kaye and Malleson could not resist casting the "Mutiny" as an existential struggle, a "noble egotism," that renewed the identity and purpose of the British nation: "it was the vehement self-assertion of the Englishman that produced this conflagration; it was the same vehement self-assertion that enabled him, by God's blessing, to trample it out" (*History of the Indian Mutiny*, I: xi). Doyle's text offers no such easily heroic interpretations, for the egotism on display throughout the narrative never aspires to nobility and "the vehement self-assertion" scarcely rises above the criminal.

*The Sign of Four* is a deeply ambivalent, and potentially conflicted, representation of empire and its most salient military and historical event to date, the Indian "Mutiny." By his own admission, Jonathan Small is a murderer, felon, thief, and possibly a rapist, and yet he remains true to the oath sworn with his Indian accomplices at knife point to the very end of the narrative; disenfranchised by the very imperial venture that was supposed to provide for him, Small's only act of social solidarity as a poor white is doubly ironic. Major Sholto, a man who does conspicuously well from his time in India, is a gambler, drunkard, thief, and may be guilty of manslaughter on more than one occasion. Despite decades in India, Captain Morstan fails to provide for his only daughter, and is more interested in securing his illegitimate share of someone's else's treasure than in seeing her; if we believe Sholto's account, he dies of a strain to his heart attempting to purloin the Agra treasure. Abel White fails to live up to his name, and proves to be arrogant and headstrong, perishing in the flames of the "Mutiny." The other British officers on the Andaman penal settlement, both civilian and military, are dissolute gamblers (and possibly drunkards) to the last man; the surgeon, Dr. Somerton, is complicit in fixing card games designed to defraud the army officers. Small joins the "Sign of Four" motivated by both his instinct for survival *and* his instinct to keep the authorities from claiming the Agra treasure; this is, from start to finish, a story of finders keepers.

Defrauding the government of tax revenue at home and abroad seems to be something in which almost everyone in *The Sign of Four* is complicit. Small joins his Indian accomplices to prevent the government claiming the Agra treasure, while both Captain Morstan and Major Sholto prefer to become participants (and potential beneficiaries) of Indian loot rather than declare it to the authorities. Bartholomew Sholto prefers to make Mary Morstan a token recompense of a single pearl each year, but neither brother declares the illegal hoard to the police. Neither Watson nor Mary Morstan seems to have any compunction in handling stolen goods, with her "share" of the treasure briefly masquerading as a putative dowry. Only Holmes and Tonga remain indifferent to the glittering appeal of the Agra treasure. Underpinning Doyle's text and the grasping individualism that it depicts, there is clearly a profound libertarian scepticism of the system of government-administered tax revenue and expenditure. This is all the more ironic as it was the alleged competence of British administration—the "rule of law"—that was the central ideology justifying direct colonial rule in India. Faced with the magnificent temptation of ill-gotten Indian treasure, *The Sign of Four* suggests that social solidarity, like Small's pithy excuse that "the rajah had been deposed ... so no one had any particular interest" (147) in the jewels, is little more than a comforting and delusional fiction. For the reader, it is a suspension of critical disbelief too far in Doyle's novel of crime and detection.

## Eradicating Savagery

Victorian ethnologists, as Patrick Brantlinger has noted, were consumed by the need to measure, define, and classify humanity through the production of an explicitly hierarchical "ethnological map," which would "help in the general process of eradicating savagery, if not the savages themselves" (Brantlinger, 175). Unsurprisingly, the Ethnological Society launched its campaign to map the population of British administered India in 1868, a process that would culminate in the 1871 national census. "Savagery" existed at home as well as abroad. If races could be identified through the newly mastered technology of photography and classified through the new science of anthropometry, so could other groups: beggars, vagabonds, degenerates, and criminals. The influential work of Lombrooo, Ellin, and Galton had already provided a methodology for the identification of a variety of "criminal" types, but what is often less evident is the extent to which these discourses depended on a larger narrative: that of race.

Late Victorian social scientists repeatedly and insistently drew links between unregenerate "primitive" groups abroad and the degenerating urban criminal underclass at home. The Scottish anthropologist Edward Burnett Tylor (1832-1917) unequivocally declared that "in our great cities, the so-called 'dangerous classes' are sunk in hideous misery and depravity"; comparing them to the Kanaks of New Caledonia, he confirmed that "we may sadly acknowledge in our midst something worse than savagery" (Tylor, I: 42-3). John Beddoe in *The Races of Britain* (1885) explicitly linked class stratification in Britain with the idea of multiple, competing races. The recidivist criminal underclass at home, like the unregenerate savage abroad, was a race apart, one that had to be either classified and reformed, or removed and eradicated. Both these groups, represented by Small and Tonga, are central to the plot of Doyle's novel. In the context of late nineteenth-century racial theory, informed by explicitly determinist ideas about the survival and extinction of races, the encounter narrative between the "civilised" and the "savage" could have only one desirable outcome: the reform or elimination of the latter by the former.

One of the most dramatic illustrations of this encounter narrative was that between the British and Andamanese (see Appendix D). Nominally under British control since 1789, the islands had effectively been abandoned after the failure of the first penal settlement in 1796. In the aftermath of the Indian "Mutiny," the islands were proposed as a suitable place for the transportation of political prisoners, and an expedition led by F.J. Mouat resulted in the establishment of what would eventually become India's largest facility for transported convicts. Islands in the Andaman archipelago were re-named after the British heroes of the "Mutiny"; it was thought that both the remoteness of the islands and the hostility of the indigenous population would ensure that convicts would never be able to escape and make their way back to India.

The rhetoric of the penal settlements explicitly subscribed to the idea of the reformation of "savagery," both Indian and Andamanese. A letter to the editor of *The Times* made the connection clear: "let them be shipped off to the Andamans ... let them make head, if they can, against savages not more bloodthirsty than themselves" (Appendix D1). Mouat rose to the brief suggested by the letter to *The Times*. "There was something poetical in the retributive justice" he wrote, "that thus rendered the crimes of an ancient race the means of reclaiming a fair and fertile tract of land from the neglect, the barbarity, and the atrocities of a more primitive, but scarcely less cruel and vindictive race, whose origin is yet involved in such a dark cloud of mystery" (Mouat, 45). In reality, the penal settlement became a breeding ground for Indian nationalist dissent; even more acutely, it became a constant source of friction between the indigenous population and the immigrant British administration and Indian convicts.

Despite the ostensibly utilitarian language of reform through labour, conditions in the settlement were desperate. In the first years, prisoners at Port Blair were kept shackled together, and no distinction was made between political prisoners and common criminals (almost all murderers). Poor nutrition, epidemics of disease (especially malaria), constant attacks by the Andamanese, and the hopelessness of a sentence without prospect of remission, seriously impacted upon the experiment's viability. In the first three months, some 87 of the 730 original convicts had been hanged for desertion. The size of the penal colony continued to increase, peaking at some 12,000 prisoners in the 1890s; through the 1870s, numbers averaged around 7,000. Notwithstanding the large scale of transportation to the Andamans, relatively little attention was paid in London.

It was the murder of the Governor General of India, Lord Mayo, by an Afridi assassin at Port Blair on 8 February 1872 (see Appendixes C4, D3, and D4) that brought an unprecedented level of scrutiny from the colonial core to bear upon one of its most neglected peripheries. Shere Ali's unpredicted murder (which clearly revived memories of the "Mutiny") meant that the rhetoric of "eradicating savagery" was again applied to the settlement with renewed vigour. A new civilian police force was recruited from the mainland to stem the rising number of escapes, which in 1872 numbered some 160; no fewer than 100 convicts had fled by sea, of whom only 49 were killed or recaptured (Small's escape by sea was the most likely way out). There was also increasing outrage in the London press that far from reforming the convicts and turning them into productive settlers, the penal settlement was letting them run wild. A report in *The Times* in February 1872 (see Appendix D4) focussed on what it saw as the drunken debauchery and regression of British and Eurasian convicts; the details of the James Devine murder case suggest the plausibility of Small's account of drinking and card games in the settlement.

In the ninety years of its existence as a penal settlement, the Andaman Islands and Agra were inextricably linked. The first superintendent at Port Blair, J.P. Walker, had previously been in charge of Agra prison, long regarded by the British as their model Indian penal establishment; many in the first consignment of prisoners shipped to the islands, like Small and his collaborators, were processed through Agra prison. Cloth for their uniforms continued to be made by Agra's inmates. The very prison that had introduced the system of self-policing (Sikh inmates replaced the wardens, who had deserted en masse at the outbreak of the "Mutiny") became the model for the penal colony produced as a direct consequence of the events of 1857. The Andaman penal colony further developed the same system of self-policing pioneered by William Muir during the "Mutiny" in Agra, and extensively used the promotion of prisoners through class gradation until they eventually became overseers or self-supporting settlers (see Appendixes D3 and D5); many of the overseers were Europeans (Small is just such a partially rehabilitated convict). Under this "ticket of leave" system, even life convicts could become self-supporters after 12 years of good behaviour, and earn a conditional pardon after 21 years. The vast majority of Europeans on the Andamans were free men employed in regulating the mass of Indian convicts; the 1873-74 report lists 45 civilian men working in the administration, 108 military men, but only six European male convicts. While the Surgeon-General for the settlement was British, his three assistants were all Indian civilians recruited from the mainland, and the dispensing clerk was a distinguished political prisoner, the former chief imam of Patna, the Mufti Ahmedullah. Small would therefore have been a statistical rarity as a white convict, and, lacking any education, an unlikely candidate to be a dispensing assistant to the chief surgeon.

There was another community on the islands: the Andamanese. They had not been consulted about the imposition of one of the world's largest gulags on their homeland, and were implacably opposed to what they rightly saw as an unwarranted encroachment. Increasingly hemmed in by colonial expansion and denied the possibilities of citizenship or social mobility, what they shared with the convicts was their status as subalterns within an imperial order. In her essay "Can the Subaltern Speak?," Gayatri Chakravorty Spivak has argued that "in the context of colonial production the subaltern has no history and cannot speak" (Spivak, 287) and has urged readers to take up "the task of *measuring silences*, whether acknowledged or unacknowledged" (Spivak, 286).

*The Sign of Four* is a narrative replete with subalterns, both literal and metaphorical, and not all of them are privileged with the power of speech. Jonathan Small is a subaltern before his arrival in India, enlisting as a foot soldier; his civilian career, as an overseer on an indigo plantation armed with a whip, requires the subjugation of an undifferentiated subaltern mass of Indian peasants. As a prisoner, Small is denied autonomy and mobility; as a fugitive from justice, he joins the swollen ranks of another subaltern group, London's urban poor. Unlike the numerous Indian *khitmutgars* populating the novel, who by themselves constitute yet another conspicuous subaltern community, Small is allowed to have his say, without interruption or coercion. The greatest unwritten subaltern narrative in *The Sign of Four*, and the one that is the most difficult to represent and interpret, is that of the only Andamanese in the whole of nineteenth-century English fiction: Tonga.

Equidistant between the Fridays of Daniel Defoe's *Robinson Crusoe* (1719) and J.M. Coetzee's *Foe* (1988), and like them, presented as chattel, Tonga's enslavement at the hands of Small buys the fugitive ex-convict his temporary freedom. Small's ownership of Tonga (for that is clearly what it is) facilitates the former's revenge; ironically, Tonga's captivity is Small's freedom, and vice versa. Tonga's autonomy and capacity for speech are systematically denied in Doyle's text. Mimicking the rhetoric of nineteenth-century race theory, both Watson and Holmes methodically dehumanize Tonga; for Watson, he is a "something black" between Small's knees, a "dark mass, which looked like a Newfoundland dog" (124), his features "deeply marked with all bestiality and cruelty," his "thick lips ... writhed back from his teeth, which grinned and chattered at us with half animal fury" (125). Watson's description of Tonga is startlingly similar to the language used by F.J. Mouat in *Adventures and Researches Among the Andaman Islanders* (1863) to describe the bodies of two Andamanese men that he had shot and killed: "their features, distorted as they appeared by the most violent passions, were too horrible for anything of human mould, and I could regard them only as the types of the most ferocious and relentless fiends" (Mouat, 255-56).

Holmes unquestioningly accepts the inaccurate and outdated information about the Andamanese offered in the spurious government gazette; this is all the more anomalous as so much of his information is up to date. Doyle's depiction of the Andamanese in *The Sign of Four* drew contemporary criticism, most notably from Andrew Lang (in the *Quarterly Review* in 1904) who pointedly commented that "the Andamanese are cruelly libelled, and have neither the malignant qualities, nor the heads like mops, nor the customs, with which they are credited by Sherlock ... [they] do not use blowpipes of poisoned arrows, and show no traces of cannibalism" (Lang, 178). "Anthropology we do not expect from Sherlock," Andrew Lang acidly observed on Holmes's unquestioning acceptance of the gazetteer, adding that he had "detected the wrong savage, and injured the character of an amiable people" (Lang, 178). It is equally telling that Holmes's and Watson's shooting of Tonga is the only time in Holmes's entire fictional career that he fires a shot in anger or self-defence, simultaneously eradicating both the savage and his alleged savagery. In this act, they replicate a standard trope of the encounter narratives between the British and the Andamanese. Andamanese resistance to encroachment, with bows and arrows rather than Tonga's fanciful blowpipe, was met with overwhelming British force. The most dramatic such encounter, the battle of Aberdeen on 14 May 1858, resulted in the deaths of up to 400 Andamanese, without loss on the British side.

While Holmes and Watson completely deny Tonga's humanity, Small, despite literally having the whip hand, calling him all the imprecations under the sun ("hell-hound," "little devil," "bloodthirsty imp"), and parading him at freak shows as "the black cannibal" (154), does not. Small's account humanises Tonga, for it glosses both the real decimation of the Andamanese from exposure to imported diseases—"he was sick to death, and had gone to a lonely place to die"—and the inevitable impact of the encounter narrative: "he took a kind of fancy to me then, and would hardly go back to his woods" (152). Nursed back to health and pacified, Tonga is a native who can never return, and Small takes advantage of his isolated vulnerability. "He was staunch and true, was little Tonga," Small declares, in an unlikely act of solidarity; "no man ever had a more faithful mate" (152). Such collaboration would have been highly unlikely, for the British often employed Andaman trackers to catch fugitive escaped convicts, and many Andamanese, especially the Jarawa on Great Andaman, remained implacably opposed to the presence of the penal settlement, attacking trespassers on sight.

Tonga is denied the ability to speak by Watson (he sees his chattering as animal and not human) and Holmes, despite being a self-confessed linguist, makes no attempt to communicate with him, but Small apparently does, for he has "learned a little of his lingo from him, and this made him all the fonder of me" (152); whether Tonga learns English is never suggested. The more than a dozen mutually unintelligible Andamanese languages proved to be a significant challenge to British administrators and anthropologists alike. By the end of the nineteenth century, only one had been the subject of detailed study (E.H. Man's *Grammar of the Bojingijîda or South Andaman language*, 1878), and even the distinguished anthropologist Alfred Radcliffe-Brown found the task of learning any Andamanese language to be beyond him (he relied upon interpreters for field work). Andamanese held in captivity by the British proved to be capable linguists, picking up

Bengali, Hindi, Tamil, Malay, Burmese, or English without difficulty. "Mary Andaman," an Andamanese girl captured by the British in 1840, grew up to become a schoolmistress in Singapore, while "Topsy," who had been Christianised at the end of the nineteenth century, learned fluent English and even travelled to London, where she met both the anthropologist E.H. Man and the former administrator of the islands, Richard Carnac Temple. Dozens of Andamanese were carried off into captivity by the British, either to confirm existing opinions of their primitive and irredeemably savage state, or else to provide physical evidence of the success of the project to "civilise" them (for an account of Mouat's parading of "Jack Andaman" in Calcutta, see Appendix D2). In the majority of these accounts, the voices of the Andamanese themselves are elided or obscured. "Topsy," Madhusree Mukerjee observes, wrote the only account of the encounter narrative between the forces of "civilisation" and "savagery" from the Andamanese perspective. Radically different from Tonga's subterfuge existence huddled "in some sort of dark ulster or blanket" (125), her account of London included journeys on the underground, visits to Westminster Abbey ("a beautiful old abbey"), Windsor Castle ("I never saw any place so beautiful"), and the British Museum (Mukerjee, 50).

If, like "Topsy," Tonga could speak, what would he say? More than likely, Tonga would offer us a sequence of events with which almost all Andamanese in the mid- and late nineteenth century would have been familiar: the traumatic arrival of thousands of outsiders on his ancestral land, who brought with them diseases to which he had no resistance and which decimated his family, the imposition of an uninvited penal settlement, whose "reformed" convict-settlers continuously encroached on his land, and finally, a captivity narrative as the chattel of an Englishman, which included his obligatory parading as a "savage" and ended with his death at their hands. Tonga's narrative, like that of his profusely voluble double, Jonathan Small, bears mute testimony to the nineteenth-century's obsession with the reform and eradication of "savagery" at home and abroad.

Tonga's death on the Thames, subtitled "The End of the Islander," is suitably metonymic, for it symbolises not just the death of an individual, but the near-extermination of a people. By the time Doyle came to write *The Sign of Four*, the vast majority of the indigenous population of the islands had been decimated by imported diseases (especially syphilis) to which they had no resistance, a problem exacerbated by the imposition of the

"Homes Policy," a system of reservations that helps spread communicable diseases. At the time the British established the penal colony in 1858, there were between five and eight thousand Andamanese living on the island of Great Andaman; the 1901 census counted only 625. When Maurice Vidal Portman, the officer responsible for the Andamanese and one of their few British champions, came to write *A History of Our Relations with the Andamanese* in the last year of the nineteenth century, he was forced to accept that the narrative of encounter between the British and the Andamanese, like Tonga's elimination, could have only one possible outcome. "So long as they were left to themselves and not in any way interfered with by outside influences," Portman wrote in a register of elegy, "they would continue to live; but when we came amongst them and admitted the air of the outside world, with consequent changes, to suit our necessities, not theirs, they lost their vitality, which was wholly dependent on being untouched, and the end of the race came" (Portman, 875).

Anticipating the extinction of the Andamanese (like that of the indigenous Tasmanians before them) as a direct consequence of "civilisation," Portman dedicated his time on the islands to relentlessly recording, documenting, and measuring the surviving population that had been "pacified." Some of Portman's photographs are remarkably beautiful and empathetic images, but many others are harrowing examples of the racial arrogance of late nineteenth-century anthropometry. The archive includes photographs of men and women with their arms forcibly extended against rulers, and children with their heads in the prongs of measuring devices. "It was the fate of the Great Andamanese," Madhusree Mukerjee poignantly observes about Portman's obsessive archive, "to be documented into their graves" (Mukerjee, 61). Containing some 11 volumes of private papers and over 300 photographs divided between London and Kolkata, the Portman collection demonstrates the extent to which an archive effectively displaces the very people it is supposed to represent.

Against the odds, and despite the British nineteenth-century belief in the "eradication of savagery" through the imposition of the benefits of civilisation, Portman's resignation to their imminent extinction, and Holmes's and Watson's active participation in Tonga's elimination, a small number of Andamanese, perhaps as few as 500 people, have survived into the twenty-first century. The question of what Andamanese readers might make of Tonga's subaltern silence in Doyle's narrative of imperialism and its discontents remains unanswered, but not, I hope, forever.

# Arthur Conan Doyle: A Brief Chronology

1859    Arthur Ignatius Conan Doyle (hereafter "ACD") born
        at 11 Picardy Place, Edinburgh, on 22 May. He is the
        third of the nine children of Charles Altamont Doyle
        (1832-93), an alcoholic draughtsman and artist, and
        Mary Foley (1838-1921), an Irish immigrant.

1864    Charles Doyle's alcoholism causes a temporary
        family break-up; ACD lives with the family of the
        historiographer-royal for Scotland, John Hill Burton.

1867    ACD's parents are reunited, and move into over-
        crowded tenement flats. ACD becomes leader of a
        local street gang.

1868    Attends Hodder preparatory school, 1868-70; paid
        for by rich uncles.

1870    Attends Stonyhurst College, England's leading Jesuit
        boarding school, from 1870-75.

1875-76 Spends his final school year at Feldkirch, Austria.
        Loses his formal belief in Catholicism.

1876    Enters Edinburgh University as a medical student;
        influenced by Joseph Bell, who specialised in deduc-
        tion from minutiae (and served as the model for
        Sherlock Holmes).

1879    "The Mystery of the Sasassa Valley," his first short
        story, is published in *Chambers's Edinburgh Journal*,
        earning him the princely sum of £3 3s.; first non-
        fiction, "Gelseminum as a poison," published in the
        *British Medical Journal* in the same month.

1880    February-September: Works as ship's doctor on the
        Greenland bound whaling ship, the *Hope*.

1881    Graduates from medical school.

1881    October-January 1882: Works as a surgeon on board
        the steamer *Mayamba* off the West African coast; his
        patients include the African-American anti-slavery
        campaigner Henry Highland Garnet.

1882    In Plymouth, before moving to Southsea (near
        Portsmouth) where he establishes a successful
        medical practice. First encounters theosophy.

1884    January: "The Captain of the *Pole-Star*" published in
        *Cornhill*.

1885    August: Marries Louisa Hawkins (1856-1906), sister of one of his patients. ACD's father Charles is by this point confined to Scottish mental institutions. Earns a doctorate from Edinburgh University for his thesis on aspects of syphilis.

1887    November: ACD's first Holmes and Watson story, "A Study in Scarlet," appears in *Beeton's Christmas Annual* (it is rejected by *Cornhill*) before being published by Ward Lock & Co. for a paltry £25.

1888    December: *The Mystery of Cloomber* published by Ward and Downey, after serialisation in the *Pall Mall Gazette*.

1889    Daughter Mary born. His most famous historical romance, *Micah Clarke* (set during the Monmouth Rebellion), is published by Longmans after a recommendation from Andrew Lang. August: together with Oscar Wilde (*The Picture of Dorian Gray*), commissioned to write *The Sign of Four* for £100 for *Lippincott's Monthly Magazine* (ACD's income from his medical practice was £300 per year).

1890    *The Sign of Four* published by Spencer Blackett after serialisation in *Lippincott's*.

1891    Runs a short-lived ophthalmology clinic in Marylebone, but the success of the first six Holmes stories in the *Strand* makes him give up medicine. Moves to 12 Tennison Road, South Norwood. *The White Company*, a three-volume historical novel set in the fourteenth century, becomes a sales success, and cements his reputation as one of the leading historical novelists of the age.

1892    Son Alleyne Kingsley Conan Doyle born.

1893    ACD's father Charles dies of alcoholism. December: struggling to satisfy his readers' insatiable demands, ACD apparently kills off both Holmes and his fictional nemesis Moriarty at the Reichenbach Falls in "The Adventure of the Final Problem" (*Strand Magazine*).

1894    Actively involved with the Society for Psychical Research. September: leaves for lecture tour of America, where he is lauded; spends Thanksgiving with Rudyard Kipling in Vermont.

1895    December-May 1896: Winters in Egypt for the sake of his wife's health.

1896    Publishes *Rodney Stone*, a historical romance about bare-knuckle boxing; Smith, Elder & Co. pay him £4,000 in advance royalties.

1897  Publishes *Uncle Bernac*, a historical romance set during the Napoleonic Wars. March: meets and immediately falls in love with Jean Blyth Leckie (1872-1940).

1898  Publication of *The Tragedy of the Korosko*.

1900  Becomes actively involved in supporting British policy in South Africa; writes *The Great Boer War*. Unsuccessfully stands for election as a Unionist candidate for an Edinburgh constituency.

1901  *The Hound of the Baskervilles* serialised in the *Strand*.

1902  *The War in South Africa: its Cause and Conduct* published. ACD is by now a vociferous and influential propagandist for the British in the Boer War. *The Hound of the Baskervilles*, effectively a Holmes and Watson prequel, is published to widespread acclaim.

1903  Offered $45,000 by the American *McClure's Magazine*, ACD revives Sherlock Holmes.

1905  *The Return of Sherlock Holmes* gathers together 13 stories serialised in the *Strand*, 1903-04.

1906  July: ACD's wife Louisa dies of tuberculosis. Unsuccessfully stands as a Unionist candidate for Parliament for Hawick and Galashiels, in the Scottish borders. Publishes *Sir Nigel*, a prequel to *The White Company*, and is paid $25,000 for serial rights in America.

1907  September: Marries Jean Leckie, whom he has already known for over a decade. Publishes *Through the Magic Door*, a personal account of the books that influenced him. Champions the wrongful conviction of George Edalji.

1908  Publication of *Round the Fire Stories*, a collection of tales of the grotesque and supernatural. Moves to his new house in Crowborough, Sussex.

1909  Son Denis born. Writes *The Crime of the Congo*, a searing indictment of Belgian rule in Central Africa.

1910  Son Adrian born.

1911  Begins espousing the cause of Home Rule for Ireland.

1912  *The Lost World*; daughter Jean Lena Annette born (she grows up to become head of the Women's Royal Air Force in World War II). Publicly quarrels with George Bernard Shaw over the sinking of the *Titanic*.

1914  The last Holmes and Watson novel, *The Valley of Fear*, serialised in the *Strand*.

1916  Serves as a military correspondent and historian, publishing a number of volumes on the progress of World War I. By now ACD is a committed spiritualist.

1917   "His Last Bow," subtitled "The War Service of Sherlock Holmes," appears in the *Strand*.
1918   October: Death of son Kingsley after being wounded at the Somme, just days before the end of World War I.
1920   *The British Campaign in France and Flanders* (6 volumes).
1921   Writes *The Wanderings of a Spiritualist*. Spends the next decade campaigning for spiritualism.
1924   Publishes his memoir *Memories and Adventures* (1924-28).
1926   *The History of Spiritualism* published.
1927   April: the sixtieth and last Holmes story, "The Adventure of Shoscombe Old Place," appears in the *Strand*. The final Holmes and Watson volume, *The Case-Book of Sherlock Holmes*, is published.
1930   7 July: ACD dies at his home, Windlesham, in Sussex.

# A Note on the Text

The first publication of *The Sign of Four* was in the pages of the Philadelphia literary journal, *Lippincott's Monthly Magazine*, in the February 1890 number, where it shared its billing with Oscar Wilde's novel of aestheticism, decadence, and murder in London, *The Picture of Dorian Gray*. In fact, both authors had been specially commissioned to write for *Lippincott's* by the magazine's agent, Joseph Stoddard, at a dinner given at the Langham Hotel on 30 August 1889. In *The Sign of Four*, the Langham is the hotel in which Captain Morstan stays on his return from India.

The copy text for this edition is the first English volume edition, published by Spencer Blackett of St. Bride Street, London, on 1 October 1890; British Library call mark 012632.h.5. Apart from silently correcting a few obvious misprints, this Broadview edition faithfully reproduces the Spencer Blackett text, including its preferences for capitalisation, punctuation, italicisation, and the single definite article in the title. The Spencer Blackett edition was set from the same plates as the *Lippincott's Magazine* February 1890 serialisation, but features 23 textual variants. Perhaps a more evident difference between the first serialisation and the first volume edition is in Spencer Blackett's far more frequent use of paragraph breaks in the text.

The frontispiece to Spencer Blackett's first edition explicitly promoted Doyle as a writer of historical novels ("The author of *Micah Clarke, The Firm of the Girdlestone* etc.") and failed to mention his first (and only previous) Sherlock Holmes story, *A Study in Scarlet* (1887). This suggests that both author and publisher were still unsure about Holmes's prospects in the marketplace (this was only the second Holmes story). However, any uncertainty was eventually dispelled by the growing circulation figures (averaging over 300,000 copies in 1891) for the Holmes stories in the *Strand*, a journal with which he soon become synonymous. Successive British editions of *The Sign of Four* were offered at lower prices and reached larger readerships; George Newnes's decision to include it as the ninth title of his "Penny Library of Famous Books" in 1896 ensured sales of over 50,000 copies. As well as cheap editions for the British market, *The Sign of Four* was extensively serialised in the popular press; it was reprinted in the *Bristol Observer*, the *Hampshire Telegraph and Sussex Chronicle*, the *Birmingham Weekly Mercury*, the *Glasgow Weekly Citizen*, and famously, in George Newnes's *Tit-Bits*. It also featured as number 2698 in Bernhard Tauchnitz's famous and immensely popular "Collection of British Authors" series (Leipzig, 1891). The history of Sherlock Holmes's incredible and enduring popularity is also, in part, the history of cheap mechanised printing. For more on this, see Richard Lancelyn Green and John Michael Gibson, *A Bibliography of A. Conan Doyle* (London: Hudson House, 1999).

## American Piracy

*The Sign of Four* was published before the passage of the first Anglo-American copyright law, the Chace Act, which came into force on 1 July 1891; the Act demanded simultaneous publication on both sides of the Atlantic, and did not offer any retrospective copyright protection. As a result, *The Sign of Four* was not protected under United States copyright law, and unauthorised American reprinting of the book gathered pace. Donald A. Redmond notes that well over 200 different editions and printings of *The Sign of Four* took place in America during Doyle's lifetime; this is a process that has continued, for the text remains, as it always has been in the USA, in the public domain. For a detailed study of the extent of American piracy of *The Sign of Four* and its implications, see Redmond's *Sherlock Holmes Among the Pirates: Copyright and Conan Doyle in America 1890-1930* (London: Greenwood Press, 1990), 25-50, 113-196 and Nathan L. Bengis, *The "Signs" of Our Times: An Irregular Bibliography of Arthur Conan Doyle's Novel "The Sign of Four"* (New York: 1956). For more on the general impact of the 1891 Copyright Act, see James L. West III, "The Chace Act and Anglo-American Literary Relations," *Studies in Bibliography: Papers of the Bibliographical Society of the University of Virginia* 45 (1992): 303-11.

## *The Sign of Four* or *The Sign of the Four*?

All subsequent British editions have followed the Spencer Blackett edition's title, *The Sign of Four*. In contrast, and caused by the sheer range of proliferating editions and the accumulation of distortions in multiple transmissions, American editions have been inconsistent: *The Sign of the Four, At the Sign of the Four, Sign of the Four, A Sign of Four, The Sign of the Four or the Problem of the Sholtos*, and *Sherlock Holmes and the Sign of the "4"* have all appeared in unauthorised texts. Arthur Conan Doyle himself usually referred to the work as *The Sign of Four*, despite the fact that the original manuscript (currently in an American private collection) carried the grammatically and semantically more precise title, *The Sign of the Four*. However, the title on the MS is not in Doyle's own hand, suggesting that this cannot be accepted as a definitive, authorial intervention. The first and all subsequent British editions consistently omit this second definite article, and this has led me to prefer *The Sign of Four* for this edition.

# The Sign of Four

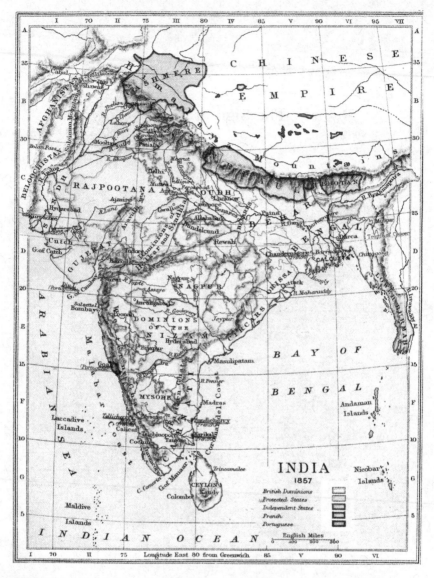

India, 1857, by Samuel Rawson Gardiner D.C.L., L.L.D., *School Atlas of English History* (London: Longmans, Green, and Co., 1914).

# Chapter 1
## The Science of Deduction

Sherlock Holmes took his bottle from the corner of the mantelpiece, and his hypodermic syringe from its neat morocco case. With his long, white, nervous fingers he adjusted the delicate needle and rolled back his left shirt-cuff. For some little time his eyes rested thoughtfully upon the sinewy forearm and wrist, all dotted and scarred with innumerable puncture-marks. Finally, he thrust the sharp point home, pressed down the tiny piston, and sank back into the velvet-lined arm-chair with a long sigh of satisfaction.

Three times a day for many months I had witnessed this performance, but custom had not reconciled my mind to it. On the contrary, from day to day I had become more irritable at the sight, and my conscience swelled nightly within me at the thought that I had lacked the courage to protest. Again and again I had registered a vow that I should deliver my soul upon the subject; but there was that in the cool, nonchalant air of my companion which made him the last man with whom one would care to take anything approaching to a liberty. His great powers, his masterly manner, and the experience which I had had of his many extraordinary qualities, all made me diffident and backward in crossing him.

Yet upon that afternoon, whether it was the Beaune[1] which I had taken with my lunch, or the additional exasperation produced by the extreme deliberation of his manner, I suddenly felt that I could hold out no longer.

"Which is it to-day," I asked, "morphine or cocaine?"[2]

He raised his eyes languidly from the old black-letter volume which he had opened.

---

1 A wine from Beaune, in the Côte d'Or department in the Burgogne (Burgundy) region of France.

2 Cocaine ($C_{27}H_{21}NO_4$) is a powerful alkaloid extracted from the leaves of the coca plant. In the 1880s, it was still viewed as a wonder drug, with the *British and Colonial Druggist* declaring on 31 July 1886 that it was a "valuable alkaloid ... whose properties as a local anaesthetic have created almost a revolution in ophthalmic and other branches of surgery." Cocaine possession and use was legal in Britain at the time, and while there is no conclusive evidence, Doyle may well have experimented with cocaine himself; he certainly would have prescribed it as a painkiller for his patients.

*so casual!*

"It is cocaine," he said, "a seven-per-cent solution. Would you care to try it?"

"No, indeed," I answered brusquely. "My constitution has not got over the Afghan[1] campaign yet. I cannot afford to throw any extra strain upon it."

He smiled at my vehemence. "Perhaps you are right, Watson," he said. "I suppose that its influence is physically a bad one. I find it, however, so transcendently stimulating and clarifying to the mind that its secondary action is a matter of small moment."

"But consider!" I said earnestly. "Count the cost! Your brain may, as you say, be roused and excited, but it is a pathological and morbid process which involves increased tissue-change and may at least leave a permanent weakness. You know, too, what a black reaction comes upon you. Surely the game is hardly worth the candle. Why should you, for a mere passing pleasure, risk the loss of those great powers with which you have been endowed? Remember that I speak not only as one comrade to another but as a medical man[2] to one for whose constitution he is to some extent answerable."

He did not seem offended. On the contrary, he put his finger-tips together, and leaned his elbows on the arms of his chair, like one who has a relish for conversation.

"My mind," he said, "rebels at stagnation. Give me problems, give me work, give me the most abstruse cryptogram, or the most intricate analysis, and I am in my own proper atmosphere. I can dispense then with artificial stimulants. But I abhor the dull routine of existence. I crave for mental exaltation. That is why I have chosen my own particular profession, or rather created it, for I am the only one in the world."

"The only unofficial detective?" I said, raising my eyebrows.

"The only unofficial consulting detective," he answered. "I am the last and highest court of appeal in detection. When Gregson, or Lestrade, or Athelney Jones are out of their depths—which, by the way, is their normal state—the matter is laid before me. I examine

---

1  Watson is a veteran of the Second Anglo-Afghan War (1878-80), during which he was injured at the Battle of Maiwand on 27 July 1880; see Appendix C.

2  Watson describes himself at the beginning of *A Study in Scarlet* as "John Watson, M.D., late of the Army Medical Department." Doyle had studied medicine at the University of Edinburgh from 1876 to 1881, and at the time of writing *The Sign of Four* was still a practising doctor.

the data, as an expert, and pronounce a specialist's opinion. I claim no credit in such cases. My name figures in no newspaper. The work itself, the pleasure of finding a field for my peculiar powers, is my highest reward. But you have yourself had some experience of my methods of work in the Jefferson·Hope[1] case."

"Yes, indeed," said I cordially. "I was never so struck by anything in my life. I even embodied it in a small brochure, with the somewhat fantastic title of 'A Study in Scarlet.'"

He shook his head sadly.

"I glanced over it," said he. "Honestly, I cannot congratulate you upon it. Detection is, or ought to be, an exact science and should be treated in the same cold and unemotional manner. You have attempted to tinge it with romanticism, which produces much the same effect as if you worked a love-story or an elopement into the fifth proposition of Euclid."[2]

"But the romance was there," I remonstrated. "I could not tamper with the facts."

"Some facts should be suppressed, or, at least, a just sense of proportion should be observed in treating them. The only point in the case which deserved mention was the curious analytical reasoning from effects to causes, by which I succeeded in unravelling it."

I was annoyed at this criticism of a work which had been specially designed to please him. I confess, too, that I was irritated by the egotism which seemed to demand that every line of my pamphlet should be devoted to his own special doings. More than once during the years that I had lived with him in Baker Street I had observed that a small vanity underlay my companion's quiet and didactic manner. I made no remark however, but sat nursing my wounded leg. I had had a Jezail[3] bullet through it some time

---

1  The subject of the first Sherlock Holmes story, *A Study in Scarlet* (1887).

2  The fifth postulate in Euclid's *Elements*, also known as the parallel postulate, states that if two lines intersect a third in such a way that the sum of the inner angles on one side is less than two right angles, then the two lines inevitably must intersect each other on that side if extended far enough.

3  A type of handmade musket widely used at the time in Afghanistan. According to the accounts offered by Doyle, Watson was injured in the battle of Maiwand, near Kandahar, in the Second Anglo-Afghan War, on 27 July 1880. In *A Study in Scarlet* (1887), Watson describes the injury as being in his shoulder.

before, and though it did not prevent me from walking it ached wearily at every change of the weather.

"My practice has extended recently to the Continent," said Holmes after awhile, filling up his old briar-root pipe. "I was consulted last week by François le Villard, who, as you probably know, has come rather to the front lately in the French detective service. He has all the Celtic power of quick intuition, but he is deficient in the wide range of exact knowledge which is essential to the higher developments of his art. The case was concerned with a will and possessed some features of interest. I was able to refer him to two parallel cases, the one at Riga in 1857, and the other at St. Louis in 1871, which have suggested to him the true solution. Here is the letter which I had this morning acknowledging my assistance."

He tossed over, as he spoke, a crumpled sheet of foreign notepaper. I glanced my eyes down it, catching a profusion of notes of admiration, with stray "magnifiques," "coup-de-maîtres" and "tours-de-force," all testifying to the ardent admiration of the Frenchman.

"He speaks as a pupil to his master," said I.

"Oh, he rates my assistance too highly," said Sherlock Holmes lightly. "He has considerable gifts himself. He possesses two out of the three qualities necessary for the ideal detective. He has the power of observation and that of deduction. He is only wanting in knowledge, and that may come in time. He is now translating my small works into French."

"Your works?"

"Oh, didn't you know?" he cried, laughing. "Yes, I have been guilty of several monographs. They are all upon technical subjects. Here, for example, is one 'Upon the Distinction between the Ashes of the Various Tobaccos.' In it I enumerate a hundred and forty forms of cigar, cigarette, and pipe tobacco, with coloured plates illustrating the difference in the ash. It is a point which is continually turning up in criminal trials, and which is sometimes of supreme importance as a clue. If you can say definitely, for example, that some murder had been done by a man who was smoking an Indian lunkah,[1] it obviously narrows your field of search. To the trained eye there is as much difference

---

1  A kind of thin, hand-rolled Indian cigar, open at both ends.

between the black ash of a Trichinopoly[1] and the white fluff of bird's-eye[2] as there is between a cabbage and a potato."

"You have an extraordinary genius for minutiae," I remarked.

"I appreciate their importance. Here is my monograph upon the tracing of footsteps, with some remarks upon the uses of plaster of Paris as a preserver of impresses. Here, too, is a curious little work upon the influence of a trade upon the form of the hand, with lithotypes of the hands of slaters, sailors, cork-cutters, compositors, weavers, and diamond-polishers. That is a matter of great practical interest to the scientific detective—especially in cases of unclaimed bodies, or in discovering the antecedents of criminals. But I weary you with my hobby."

"Not at all," I answered earnestly. "It is of the greatest interest to me, especially since I have had the opportunity of observing your practical application of it. But you spoke just now of observation and deduction. Surely the one to some extent implies the other."

"Why, hardly," he answered, leaning back luxuriously in his armchair and sending up thick blue wreaths from his pipe. "For example, observation shows me that you have been to the Wigmore Street[3] Post-Office this morning, but deduction lets me know that when there you dispatched a telegram."

"Right!" said I. "Right on both points! But I confess that I don't see how you arrived at it. It was a sudden impulse upon my part, and I have mentioned it to no one."

"It is simplicity itself," he remarked, chuckling at my surprise—"so absurdly simple that an explanation is superfluous; and yet it may serve to define the limits of observation and of deduction. Observation tells me that you have a little reddish mould adhering to your instep. Just opposite the Wigmore Street Office they have taken up the pavement and thrown up some earth, which lies in such a way that it is difficult to avoid treading in it in entering. The earth is of this peculiar reddish tint which is found, as far as I know, nowhere else in the neighbourhood. So much is observation. The rest is deduction."

---

1  A type of cigar made in the city of Trichinopoly, now called Tiruchirappalli, a major city on the right bank of the Kaveri River, some 316 kilometres (195 miles) southwest of Chennai.

2  Pipe tobacco that has been cut into segments.

3  A street in Marylebone that runs parallel to Oxford Street; it is a short walk from 221b Baker Street.

"How, then, did you deduce the telegram?"

"Why, of course I knew that you had not written a letter, since I sat opposite to you all morning. I see also in your open desk there that you have a sheet of stamps and a thick bundle of post-cards. What could you go into the post-office for, then, but to send a wire? Eliminate all other factors, and the one which remains must be the truth."

"In this case it certainly is so," I replied after a little thought. "The thing, however, is, as you say, of the simplest. Would you think me impertinent if I were to put your theories to a more severe test?"

"On the contrary," he answered, "it would prevent me from taking a second dose of cocaine. I should be delighted to look into any problem which you might submit to me."

"I have heard you say it is difficult for a man to have any object in daily use without leaving the impress of his individuality upon it in such a way that a trained observer might read it. Now, I have here a watch which has recently come into my possession. Would you have the kindness to let me have an opinion upon the character or habits of the late owner?"

I handed him over the watch with some slight feeling of amusement in my heart, for the test was, as I thought, an impossible one, and I intended it as a lesson against the somewhat dogmatic tone which he occasionally assumed. He balanced the watch in his hand, gazed hard at the dial, opened the back, and examined the works, first with his naked eyes and then with a powerful convex lens. I could hardly keep from smiling at his crestfallen face when he finally snapped the case to and handed it back.

"There are hardly any data," he remarked. "The watch has been recently cleaned, which robs me of my most suggestive facts."

"You are right," I answered. "It was cleaned before being sent to me."

In my heart I accused my companion of putting forward a most lame and impotent excuse to cover his failure. What data could he expect from an uncleaned watch?

"Though unsatisfactory, my research has not been entirely barren," he observed, staring up at the ceiling with dreamy, lack-lustre eyes. "Subject to your correction, I should judge that the watch belonged to your elder brother, who inherited it from your father."

"That you gather, no doubt, from the H.W. upon the back?"

"Quite so. The W. suggests your own name. The date of the watch is nearly fifty years back, and the initials are as old as the watch: so it was made for the last generation. Jewellery usually descends to the eldest son, and he is most likely to have the same name as the father. Your father has, if I remember right, been dead many years. It has, therefore, been in the hands of your eldest brother."

"Right, so far," said I. "Anything else?"

"He was a man of untidy habits—very untidy and careless. He was left with good prospects, but he threw away his chances, lived for some time in poverty with occasional short intervals of prosperity, and finally, taking to drink, he died. That is all I can gather."

I sprang from my chair and limped impatiently about the room with considerable bitterness in my heart.

"This is unworthy of you, Holmes," I said. "I could not have believed that you would have descended to this. You have made inquiries into the history of my unhappy brother, and you now pretend to deduce this knowledge in some fanciful way. You cannot expect me to believe that you have read all this from his old watch! It is unkind and, to speak plainly, has a touch of charlatanism in it."

"My dear doctor," said he kindly, "pray accept my apologies. Viewing the matter as an abstract problem, I had forgotten how personal and painful a thing it might be to you. I assure you, however, that I never even knew that you had a brother until you handed me the watch."

"Then how in the name of all that is wonderful did you get these facts? They are absolutely correct in every particular."

"Ah, that is good luck. I could only say what was the balance of probability. I did not at all expect to be so accurate."

"But it was not mere guess-work?"

"No, no: I never guess. It is a shocking habit—destructive to the logical faculty. What seems strange to you is only so because you do not follow my train of thought or observe the small facts upon which large inferences may depend. For example, I began by stating that your brother was careless. When you observe the lower part of that watch-case you notice that it is not only dinted in two places but it is cut and marked all over from the habit of keeping other hard objects, such as coins or keys, in the same pocket. Surely it is no great feat to assume that a man who treats a fifty-guinea[1] watch so cavalierly must be a careless man. Neither is it a very far-fetched inference that a man who inherits one article of such value is pretty well provided for in other respects."

I nodded to show that I followed his reasoning.

"It is very customary for pawnbrokers in England, when they take a watch, to scratch the numbers of the ticket with a pin-point upon the inside of the case. It is more handy than a label, as there is no risk of the number being lost or transposed. There are no less than four such numbers visible to my lens on the inside of this case. Inference—that your brother was often at low water. Secondary inference—that he had occasional bursts of prosperity, or he could not have redeemed the pledge. Finally, I ask you to look at the inner plate, which contains the keyhole. Look at the thousands of scratches all round the hole—marks where the key has slipped. What sober man's key could have scored those grooves? But you will never see a drunkard's watch without them. He winds it at night, and he leaves these traces of his unsteady hand. Where is the mystery in all this?"

"It is as clear as daylight," I answered. "I regret the injustice which I did you. I should have had more faith in your marvellous faculty. May I ask whether you have any professional inquiry on foot at present?"

"None. Hence the cocaine. I cannot live without brain-work. What else is there to live for? Stand at the window here. Was ever such a dreary, dismal, unprofitable world? See how the yellow fog swirls down the street and drifts across the dun-coloured houses. What could be more hopelessly prosaic and material? What is the use of having powers, doctor, when one has no field upon which to exert them? Crime is commonplace, existence is commonplace, and no qualities save those which are commonplace have any function upon earth."

I had opened my mouth to reply to this tirade when, with a crisp knock, our landlady entered, bearing a card upon the brass salver.

"A young lady for you, sir," she said, addressing my companion.

"Miss Mary Morstan," he read. "Hum! I have no recollection of the name. Ask the young lady to step up, Mrs. Hudson. Don't go, Doctor. I should prefer that you remain."

---

1 A guinea was worth £1 1s (or 21s), so 50 guineas would be £52 10s. In today's money, this watch would be in the region of £2,600, or $5,200.

## Chapter 2
## The Statement of the Case

Miss Morstan entered the room with a firm step and an outward composure of manner. She was a blonde young lady, small, dainty, well gloved, and dressed in the most perfect taste. There was, however, a plainness and simplicity about her costume which bore with it a suggestion of limited means. The dress was a sombre grayish beige, untrimmed and unbraided, and she wore a small turban of the same dull hue, relieved only by a suspicion of white feather in the side. Her face had neither regularity of feature nor beauty of complexion, but her expression was sweet and amiable, and her large blue eyes were singularly spiritual and sympathetic. In an experience of women which extends over many nations and three separate continents, I have never looked upon a face which gave a clearer promise of a refined and sensitive nature. I could not but observe that as she took the seat which Sherlock Holmes placed for her, her lip trembled, her hand quivered, and she showed every sign of intense inward agitation.

"I have come to you, Mr. Holmes," she said, "because you once enabled my employer, Mrs. Cecil Forrester, to unravel a little domestic complication. She was much impressed by your kindness and skill."

"Mrs. Cecil Forrester," he repeated thoughtfully. "I believe that I was of some slight service to her. The case, however, as I remember it, was a very simple one."

"She did not think so. But at least you cannot say the same of mine. I can hardly imagine anything more strange, more utterly inexplicable, than the situation in which I find myself."

Holmes rubbed his hands, and his eyes glistened. He leaned forward in his chair with an expression of extraordinary concentration upon his clear-cut, hawk-like features.

"State your case," said he, in brisk, business tones.

I felt that my position was an embarrassing one.

"You will, I am sure, excuse me," I said, rising from my chair.

To my surprise, the young lady held up her gloved hand to detain me.

"If your friend," she said, "would be good enough to stop, he might be of inestimable service to me."

I relapsed into my chair.

"Briefly," she continued, "the facts are these. My father was an officer in an Indian regiment, who sent me home when I was quite a child. My mother was dead, and I had no relative in England. I was placed, however, in a comfortable boarding establishment at Edinburgh, and there I remained until I was seventeen years of age. In the year 1878 my father, who was senior captain of his regiment, obtained twelve months' leave and came home. He telegraphed to me from London that he had arrived all safe and directed me to come down at once, giving the Langham Hotel[1] as his address. His message, as I remember, was full of kindness and love. On reaching London I drove to the Langham, and was informed that Captain Morstan was staying there, but that he had gone out the night before and had not returned. I waited all day without news of him. That night, on the advice of the manager of the hotel, I communicated with the police, and next morning we advertised in all the papers. Our inquiries led to no result; and from that day to this no word has ever been heard of my unfortunate father. He came home with his heart full of hope to find some peace, some comfort, and instead—"

She put her hand to her throat, and a choking sob cut short the sentence.

"The date?" asked Holmes, opening his note-book.

"He disappeared upon the 3rd of December, 1878—nearly ten years ago."

"His luggage?"

"Remained at the hotel. There was nothing in it to suggest a clue—some clothes, some books, and a considerable number of curiosities from the Andaman Islands. He had been one of the officers in charge of the convict-guard there."

"Had he any friends in town?"

"Only one that we know of—Major Sholto, of his own regiment, the 34th Bombay Infantry.[2] The major had retired some little time before and lived at Upper Norwood.[3] We communicated with him, of course, but he did not even know that his brother officer was in England."

"A singular case," remarked Holmes.

---

1 Opened in 1865 at 1C Portland Place, one of the finest hotels in London, and one of the first (in 1879) to install electric lighting.

2 An imaginary unit.

3 A fairly respectable southeast London suburb, with the postal code SE19, Upper Norwood occupies one of the highest points in London and has commanding views of the city. Arthur Conan Doyle himself moved to Tennison Road, South Norwood, in 1891, and lived there until 1894. He used the area for the setting of the Holmes short story, "The Adventure of the Norwood Builder" (1903).

"I have not yet described to you the most singular part. About six years ago—to be exact, upon the 4th of May, 1882—an advertisement appeared in the *Times* asking for the address of Miss Mary Morstan, and stating that it would be to her advantage to come forward. There was no name or address appended. I had at that time[1] just entered the family of Mrs. Cecil Forrester in the capacity of governess. By her advice I published my address in the advertisement column. The same day there arrived through the post a small cardboard box addressed to me, which I found to contain a very large and lustrous pearl. No word of writing was enclosed. Since then every year upon the same date there has always appeared a similar box, containing a similar pearl, without any clue as to the sender. They have been pronounced by an expert to be of a rare variety and of considerable value. You can see for yourself that they are very handsome."

She opened a flat box as she spoke and showed me six of the finest pearls that I had ever seen.

"Your statement is most interesting," said Sherlock Holmes. "Has anything else occurred to you?"

"Yes, and no later than to-day. That is why I have come to you. This morning I received this letter, which you will perhaps read for yourself."

"Thank you," said Holmes. "The envelope, too, please. Postmark, London, S.W. Date, July 7.[2] Hum! Man's thumb-mark on corner—probably postman. Best quality paper. Envelopes at sixpence a packet. Particular man in his stationery. No address. 'Be at the third pillar from the left outside the Lyceum Theatre[3] tonight at seven o'clock. If you are distrustful bring two friends. You are a wronged woman and shall have justice. Do not bring police. If you do, all will be in vain. Your unknown friend.' Well, really, this is a very pretty little mystery! What do you intend to do, Miss Morstan?"

"That is exactly what I want to ask you."

---

1 If Mary Morstan is 17 at the time of her father's disappearance in 1878, it indicates that she was born in 1861. She would therefore have reached 21 (the legal age of adulthood at the time) in 1882, and without an inheritance, would have no choice but to take up paid employment.

2 One of the many chronological inconsistencies in the narrative; for a fuller discussion, see the Introduction.

3 Functioning since 1765, the Lyceum Theatre (official name: Theatre Royal Lyceum and English Opera House) is located on Wellington Street (just off the Strand) in Covent Garden, the heart of London's theatre district.

"Then we shall most certainly go—you and I and—yes, why Dr. Watson is the very man. Your correspondent says two friends. He and I have worked together before."

"But would he come?" she asked, with something appealing in her voice and expression.

"I shall be proud and happy," said I, fervently, "if I can be of any service."

"You are both very kind," she answered. "I have led a retired life and have no friends whom I could appeal to. If I am here at six it will do, I suppose?"

"You must not be later," said Holmes. "There is one other point, however. Is this handwriting the same as that upon the pearl-box addresses?"

"I have them here," she answered, producing half a dozen pieces of paper.

"You are certainly a model client. You have the correct intuition. Let us see, now."

He spread out the papers upon the table, and gave little darting glances from one to the other. "They are disguised hands, except the letter," he said presently; "but there can be no question as to the authorship. See how the irrepressible Greek $e$[1] will break out, and see the twirl of the final $s$. They are undoubtedly by the same person. I should not like to suggest false hopes, Miss Morstan, but is there any resemblance between this hand and that of your father?"

"Nothing could be more unlike."

"I expected to hear you say so. We shall look out for you, then, at six. Pray allow me to keep the papers. I may look into the matter before then. It is only half-past three.[2] *Au revoir*, then."

"*Au revoir*," said our visitor; and with a bright, kindly glance from one to the other of us, she replaced her pearl-box in her bosom and hurried away.

Standing at the window, I watched her walking briskly down the street, until the gray turban and white feather were but a speck in the sombre crowd.

"What a very attractive woman!" I exclaimed, turning to my companion.

He had lit his pipe again, and was leaning back with drooping eyelids. "Is she?" he said languidly; "I did not observe."

---

1  I.e., $\epsilon$ (epsilon), the Greek lower case letter "e."

2  Yet more evidence for the improbability of the yellow, swirling fog that Holmes observes only minutes earlier.

"You really are an automaton—a calculating machine," I cried. "There is something positively inhuman in you at times."

He smiled gently.

"It is of the first importance," he cried, "not to allow your judgment to be biased by personal qualities. A client is to me a mere unit, a factor in a problem. The emotional qualities are antagonistic to clear reasoning. I assure you that the most winning woman I ever knew was hanged for poisoning three little children for their insurance-money, and the most repellent man of my acquaintance is a philanthropist who has spent nearly a quarter of a million upon the London poor."

"In this case, however—"

"I never make exceptions. An exception disproves the rule. Have you ever had occasion to study character in handwriting?[1] What do you make of this fellow's scribble?"

"It is legible and regular," I answered. "A man of business habits and some force of character."

Holmes shook his head.

"Look at his long letters," he said. "They hardly rise above the common herd. That *d* might be an *a*, and that *l* an *e*. Men of character always differentiate their long letters, however illegibly they may write. There is vacillation in his *k*'s and self-esteem in his capitals. I am going out now. I have some few references to make. Let me recommend this book—one of the most remarkable ever penned. It is Winwood Reade's 'Martyrdom of Man.'[2] I shall be back in an hour."

I sat in the window with the volume in my hand, but my thoughts were far from the daring speculations of the writer. My mind ran upon our late visitor—her smiles, the deep rich tones of her voice, the strange mystery which overhung her life. If she were seventeen at the time of her father's disappearance she must be seven-and-twenty now—a sweet age, when youth has lost its self-consciousness and become a little sobered by experience. So I sat and mused, until such dangerous thoughts came into my

---

1 Graphology, the study of handwriting, was a relatively new pseudo-science at this time. The major figure in nineteenth-century graphology was the French Catholic priest and educator, Jean-Hippolyte Michon (1806-81).

2 First published in 1872, *The Martyrdom of Man*, a secular history of western civilisation utilising the theory of evolution, was written by the historian and explorer, William Winwood Reade (1838-75). The book had already gone through eleven editions by 1886.

head that I hurried away to my desk and plunged furiously into the latest treatise upon pathology. What was I, an army surgeon with a weak leg and a weaker banking account, that I should dare to think of such things? She was a unit, a factor—nothing more. If my future were black, it was better surely to face it like a man than to attempt to brighten it by mere will-o'-the-wisps of the imagination.

## Chapter 3
### In Quest of a Solution

It was half-past five before Holmes returned. He was bright, eager, and in excellent spirits, a mood which in his case alternated with fits of the blackest depression.

"There is no great mystery in this matter," he said, taking the cup of tea which I had poured out for him; "the facts appear to admit of only one explanation."

"What! you have solved it already?"

"Well, that would be too much to say. I have discovered a suggestive fact, that is all. It is, however, *very* suggestive. The details are still to be added. I have just found, on consulting the back files of the *Times*, that Major Sholto, of Upper Norwood, late of the 34th Bombay Infantry, died upon the 28th of April, 1882."

"I may be very obtuse, Holmes, but I fail to see what this suggests."

"No? You surprise me. Look at it in this way, then. Captain Morstan disappears. The only person in London whom he could have visited is Major Sholto. Major Sholto denies having heard that he was in London. Four years later Sholto dies. *Within a week of his death* Captain Morstan's daughter receives a valuable present, which is repeated from year to year, and now culminates in a letter which describes her as a wronged woman. What wrong can it refer to except this deprivation of her father? And why should the presents begin immediately after Sholto's death unless it is that Sholto's heir knows something of the mystery and desires to make compensation? Have you any alternative theory which will meet the facts?"

"But what a strange compensation! And how strangely made! Why, too, should he write a letter now, rather than six years ago? Again, the letter speaks of giving her justice. What justice can she have? It is too much to suppose that her father is still alive. There is no other injustice in her case that you know of."

"There are difficulties; there are certainly difficulties," said Sherlock Holmes pensively; "but our expedition of to-night will solve them all. Ah, here is a four-wheeler,[1] and Miss Morstan is inside. Are you all ready? Then we had better go down, for it is a little past the hour."

I picked up my hat and my heaviest stick, but I observed that Holmes took his revolver from his drawer and slipped it into his pocket. It was clear that he thought that our night's work might be a serious one.

Miss Morstan was muffled in a dark cloak, and her sensitive face was composed but pale. She must have been more than woman if she did not feel some uneasiness at the strange enterprise upon which we were embarking, yet her self-control was perfect, and she readily answered the few additional questions which Sherlock Holmes put to her.

"Major Sholto was a very particular friend of papa's," she said. "His letters were full of allusions to the Major. He and papa were in command of the troops at the Andaman Islands, so they were thrown a great deal together. By the way, a curious paper was found in papa's desk which no one could understand. I don't suppose that it is of the slightest importance, but I thought you might care to see it, so I brought it with me. It is here."

Holmes unfolded the paper carefully and smoothed it out upon his knee. He then very methodically examined it all over with his double lens.

"It is paper of native Indian manufacture," he remarked. "It has at some time been pinned to a board. The diagram upon it appears to be a plan of part of a large building with numerous halls, corridors, and passages. At one point is a small cross done in red ink, and above it is '3.37 from left,' in faded pencil-writing. In the left-hand corner is a curious hieroglyphic like four crosses in a line with their arms touching. Beside it is written, in very rough and coarse characters, 'The sign of the four—Jonathan Small, Mahomet Singh,[2] Abdullah Khan, Dost Akbar.' No, I confess that I do not see how this bears upon the matter. Yet it is

---

1 A horse-drawn hackney carriage with four wheels typically designed to carry four passengers.

2 As "Mahomet" (correctly, Muhammad) is an Islamic name, and "Singh" a common Sikh title, this is an unlikely conflation. Doyle used similarly unlikely compound names in his earlier tale of Indian revenge, *The Mystery of Cloomber* (1889). "Dost Akbar" means "great friend" or "friend of the great," a particularly ironic name given the role he plays in the murder of Achmet.

evidently a document of importance. It has been kept carefully in a pocket-book, for the one side is as clean as the other."

"It was in his pocket-book that we found it."

"Preserve it carefully, then, Miss Morstan, for it may prove to be of use to us. I begin to suspect that this matter may turn out to be much deeper and more subtle than I at first supposed. I must reconsider my ideas."

He leaned back in the cab, and I could see by his drawn brow and his vacant eye that he was thinking intently. Miss Morstan and I chatted in an undertone about our present expedition and its possible outcome, but our companion maintained his impenetrable reserve until the end of our journey.

It was a September[1] evening and not yet seven o'clock, but the day had been a dreary one, and a dense drizzly fog lay low upon the great city. Mud-coloured clouds drooped sadly over the muddy streets. Down the Strand[2] the lamps were but misty splotches of diffused light which threw a feeble circular glimmer upon the slimy pavement. The yellow glare from the shop-windows streamed out into the steamy, vaporous air and threw a murky, shifting radiance across the crowded thoroughfare. There was, to my mind, something eerie and ghostlike in the endless procession of faces which flitted across these narrow bars of light—sad faces and glad, haggard and merry. Like all humankind, they flitted from the gloom into the light and so back into the gloom once more. I am not subject to impressions, but the dull, heavy evening, with the strange business upon which we were engaged, combined to make me nervous and depressed. I could see from Miss Morstan's manner that she was suffering from the same feeling. Holmes alone could rise superior to petty influences. He held his open notebook upon his knee, and from time to time he jotted down figures and memoranda in the light of his pocket-lantern.

At the Lyceum Theatre the crowds were already thick at the side-entrances. In front a continuous stream of hansoms[3] and

---

1  Despite repeated chances for revision, and Doyle's own awareness of this mistake, this glaring inconsistency in the chronology of the tale has remained unchanged in subsequent editions. Doyle wrote the story in September 1889, and so clearly real time impinged upon fictional time. See the Introduction for a fuller discussion.

2  A major London thoroughfare running east from Trafalgar Square to Temple Bar and the hub of Victorian theatre and nightlife.

3  A two-wheeled carriage drawn by a single horse and normally designed for two passengers, the hansom cab was patented by Joseph Hansom in

four-wheelers were rattling up, discharging their cargoes of shirt-fronted men and be-shawled, be-diamonded women. We had hardly reached the third pillar, which was our rendezvous, before a small, dark, brisk man in the dress of a coachman accosted us.

"Are you the parties who come with Miss Morstan?" he asked.

"I am Miss Morstan, and these two gentlemen are my friends," said she.

He bent a pair of wonderfully penetrating and questioning eyes upon us.

"You will excuse me, miss," he said with a certain dogged manner, "but I was to ask you to give me your word that neither of your companions is a police-officer."

"I give you my word on that," she answered.

He gave a shrill whistle, on which a street Arab[1] led across a four-wheeler and opened the door. The man who had addressed us mounted to the box, while we took our places inside. We had hardly done so before the driver whipped up his horse, and we plunged away at a furious pace through the foggy streets.

The situation was a curious one. We were driving to an unknown place, on an unknown errand. Yet our invitation was either a complete hoax—which was an inconceivable hypothesis—or else we had good reason to think that important issues might hang upon our journey. Miss Morstan's demeanour was as resolute and collected as ever. I endeavoured to cheer and amuse her by reminiscences of my adventures in Afghanistan; but, to tell the truth, I was myself so excited at our situation, and so curious as to our destination, that my stories were slightly involved. To this day she declares that I told her one moving anecdote as to how a musket looked into my tent at the dead of night, and how I fired a double-barrelled tiger cub at it. At first I had some idea as to the direction in which we were driving; but soon, what with our pace, the fog, and my own limited knowledge of London, I lost my bearings, and knew nothing, save that we seemed to be going a very long way. Sherlock Holmes was never at fault, however, and he muttered the names as the cab rattled through squares and in and out by tortuous by-streets.

---

1834, and widely used in London until the 1920s. The driver sat on a sprung seat behind the vehicle.

1  I.e., an urchin. Victorian London's large transient street population (especially its children) were the subject of intense scrutiny and anxiety in this period; see, for example, Arthur Morrison's *A Child of the Jago* (1896).

"Rochester Row,"[1] said he. "Now Vincent Square.[2] Now we come out on the Vauxhall Bridge Road.[3] We are making for the Surrey side apparently. Yes, I thought so. Now we are on the bridge. You can catch glimpses of the river."

We did indeed get a fleeting view of a stretch of the Thames, with the lamps shining upon the broad, silent water; but our cab dashed on and was soon involved in a labyrinth of streets upon the other side.

"Wordsworth Road,"[4] said my companion. "Priory Road. Lark Hall Lane. Stockwell Place. Robert Street. Cold Harbour Lane.[5] Our quest does not appear to take us to very fashionable regions."

We had indeed reached a questionable and forbidding neighbourhood. Long lines of dull brick houses were only relieved by the coarse glare and tawdry brilliancy of public-houses at the corner. Then came rows of two-storied villas, each with a fronting of miniature garden, and then again interminable lines of new, staring brick buildings—the monster tentacles which the giant city was throwing out into the country. At last the cab drew up at the third house in a new terrace. None of the other houses were inhabited, and that at which we stopped was as dark as its neighbours, save for a single glimmer in the kitchen-window. On our knocking, however, the door was instantly thrown open by a Hindoo servant, clad in a yellow turban, white loose-fitting clothes, and a yellow sash.[6] There was something strangely incongruous in this Oriental figure framed in the commonplace doorway of a third-rate suburban dwelling-house.

"The Sahib awaits you," said he, and even as he spoke, there came a high, piping voice from some inner room.

---

1 A street in Pimlico, not far from Victoria train station.
2 In Pimlico, directly adjacent to Rochester Row.
3 Just off Vincent Square, Vauxhall Bridge Road runs southeast from Victoria station through Pimlico to Vauxhall Bridge. The journey correctly traced by Holmes is circuitous.
4 Almost certainly an uncorrected error, as it should logically be Wandsworth Road, which is directly west of Vauxhall Bridge. There is no Wordsworth Road in the vicinity.
5 All streets leading from Vauxhall to Brixton.
6 The cummerbund (from the Persian and Hindi word for a sash fastened around the waist) was a standard way of denoting the highest ranked servant, almost always the head butler.

"Show them in to me, khitmutgar,"[1] it said. "Show them straight in to me."

## Chapter 4
## The Story of the Bald-Headed Man

We followed the Indian down a sordid and common passage, ill-lit and worse furnished, until he came to a door upon the right, which he threw open. A blaze of yellow light streamed out upon us, and in the centre of the glare there stood a small man with a very high head, a bristle of red hair all round the fringe of it, and a bald, shining scalp which shot out from among it like a mountain-peak from fir-trees. He writhed his hands together as he stood, and his features were in a perpetual jerk—now smiling, now scowling, but never for an instant in repose. Nature had given him a pendulous lip, and a too visible line of yellow and irregular teeth, which he strove feebly to conceal by constantly passing his hand over the lower part of his face. In spite of his obtrusive baldness he gave the impression of youth. In point of fact, he had just turned his thirtieth year.[2]

"Your servant, Miss Morstan," he kept repeating in a thin, high voice. "Your servant, gentlemen. Pray step into my little sanctum. A small place, miss, but furnished to my own liking. An oasis of art in the howling desert of South London."

We were all astonished by the appearance of the apartment into which he invited us. In that sorry house it looked as out of place as a diamond of the first water in a setting of brass. The richest and glossiest of curtains and tapestries draped the walls, looped back here and there to expose some richly-mounted painting or Oriental vase. The carpet was of amber and black, so soft and so thick that the foot sank pleasantly into it, as into a bed of moss. Two great tiger-skins thrown athwart it increased the suggestion of Eastern luxury, as did a huge hookah[3] which stood upon a mat in the corner. A lamp in the fashion of a silver dove was hung from an almost invisible golden wire in the centre of the room. As it burned it filled the air with a subtle and aromatic odour.

---

1 Persian and Hindi, butler.
2 I.e., he was born in 1858.
3 From the Arabic word for casket or bowl, a free-standing water pipe filled with perfumed tobacco.

"Mr. Thaddeus Sholto," said the little man, still jerking and smiling. "That is my name. You are Miss Morstan, of course. And these gentlemen—"

"This is Mr. Sherlock Holmes, and this Dr. Watson."

"A doctor, eh?" cried he, much excited. "Have you your stethoscope? Might I ask you—would you have the kindness? I have grave doubts as to my mitral[1] valve, if you would be so very good. The aortic[2] I may rely upon, but I should value your opinion upon the mitral."

I listened to his heart, as requested, but was unable to find anything amiss, save, indeed, that he was in an ecstasy of fear, for he shivered from head to foot.

"It appears to be normal," I said. "You have no cause for uneasiness."

"You will excuse my anxiety, Miss Morstan," he remarked airily. "I am a great sufferer, and I have long had suspicions as to that valve. I am delighted to hear that they are unwarranted. Had your father, Miss Morstan, refrained from throwing a strain upon his heart, he might have been alive now."

I could have struck the man across the face, so hot was I at this callous and offhand reference to so delicate a matter. Miss Morstan sat down, and her face grew white to the lips.

"I knew in my heart that he was dead," said she.

"I can give you every information," said he; "and, what is more, I can do you justice; and I will, too, whatever Brother Bartholomew may say. I am so glad to have your friends here, not only as an escort to you but also as witnesses to what I am about to do and say. The three of us can show a bold front to Brother Bartholomew. But let us have no outsiders—no police or officials. We can settle everything satisfactorily among ourselves without any interference. Nothing would annoy Brother Bartholomew more than any publicity."

He sat down upon a low settee[3] and blinked at us inquiringly with his weak, watery blue eyes.

"For my part," said Holmes, "whatever you may choose to say will go no further."

---

1  Valve separating the left atrium from the left ventricle of the heart.
2  Valve in the heart separating the left ventricle from the arterial system, specifically the aorta.
3  A seat for more than one person, with a back, arms, and a division of seats, and a curiously Western piece of furniture in Sholto's otherwise completely Oriental interior.

I nodded to show my agreement.

"That is well! That is well" said he. "May I offer you a glass of Chianti,[1] Miss Morstan? Or of Tokay?[2] I keep no other wines. Shall I open a flask? No? Well, then, I trust that you have no objection to tobacco-smoke, to the balsamic odour of the Eastern tobacco. I am a little nervous, and I find my hookah an invaluable sedative."

He applied a taper to the great bowl, and the smoke bubbled merrily through the rose-water. We sat all three in a semicircle, with our heads advanced and our chins upon our hands, while the strange, jerky little fellow, with his high, shining head, puffed uneasily in the centre.

"When I first determined to make this communication to you," said he, "I might have given you my address; but I feared that you might disregard my request and bring unpleasant people with you. I took the liberty, therefore, of making an appointment in such a way that my man Williams might be able to see you first. I have complete confidence in his discretion, and he had orders, if he were dissatisfied, to proceed no further in the matter. You will excuse these precautions, but I am a man of somewhat retiring, and I might even say refined, tastes, and there is nothing more unaesthetic than a policeman. I have a natural shrinking from all forms of rough materialism. I seldom come in contact with the rough crowd. I live, as you see, with some little atmosphere of elegance around me. I may call myself a patron of the arts. It is my weakness. The landscape is a genuine Corot,[3] and though a connoisseur might perhaps throw a doubt upon that Salvator Rosa,[4] there cannot be the least

---

1 Red wine produced in the hills of Chianti, south of Florence in Tuscany, and traditionally made only from the Sangiovese grape; then as now, both the area and the wine were favourites of the British.

2 Also spelled Tokaji, the most valued Hungarian wine appellation; grown mainly from the Furmint grape and produced in the region of Tokaj in northeastern Hungary, at the confluence of the Bodrog and Tisza rivers. Golden in colour and honey-like in sweetness, the most expensive variety (*eszencia*) was believed to have medicinal and restorative properties. Holmes unabashedly prefers French wines.

3 Jean-Baptiste Camille Corot (1796-1875), French landscape painter who combined neo-Classicism with early experiments in Impressionism. Corot's canvases commanded high prices; Sholto's possession of a genuine Corot once again suggests his family's considerable wealth.

4 Salvator Rosa (1615-73), Italian Baroque painter famous for his wildly expressive and romantic landscapes.

question about the Bouguereau,[1] I am partial to the modern French school."

"You will excuse me, Mr. Sholto," said Miss Morstan, "but I am here at your request to learn something which you desire to tell me. It is very late, and I should desire the interview to be as short as possible."

"At the best it must take some time," he answered; "for we shall certainly have to go to Norwood and see Brother Bartholomew. We shall all go and try if we can get the better of Brother Bartholomew. He is very angry with me for taking the course which has seemed right to me. I had quite high words with him last night. You cannot imagine what a terrible fellow he is when he is angry."

"If we are to go to Norwood, it would perhaps be as well to start at once," I ventured to remark.

He laughed until his ears were quite red.

"That would hardly do," he cried. "I don't know what he would say if I brought you in that sudden way. No, I must prepare you by showing you how we all stand to each other. In the first place, I must tell you that there are several points in the story of which I am myself ignorant. I can only lay the facts before you as far as I know them myself.

"My father was, as you may have guessed, Major John Sholto, once of the Indian Army. He retired some eleven years ago, and came to live at Pondicherry[2] Lodge in Upper Norwood. He had prospered in India, and brought back with him a considerable sum of money, a large collection of valuable curiosities, and a staff of native servants. With these advantages he bought himself a house, and lived in great luxury. My twin-brother Bartholomew and I were the only children.

"I very well remember the sensation which was caused by the disappearance of Captain Morstan. We read the details in the

---

1 William-Adolphe Bouguereau (1825-1905), realist Académie Française painter whose works were widely collected by the nouveaux riches (especially Americans); his reputation suffered a decline after World War I. The three paintings suggest the incoherence, eccentricity, and expense of Thaddeus Sholto's collection.

2 Now called Puducherry, this was a French enclave in southern India from 1673 until its incorporation as a Union Territory in the Republic of India in 1954. Following the Treaty of Cession (1956), French remained an official state language in the Union Territory of Puducherry. Throughout the British colonial period, Pondicherry (as it was then) was a refuge for Indian nationalist activists and thinkers.

papers, and knowing that he had been a friend of our father's, we discussed the case freely in his presence. He used to join in our speculations as to what could have happened. Never for an instant did we suspect that he had the whole secret hidden in his own breast, that of all men he alone knew the fate of Arthur Morstan.

"We did know, however, that some mystery, some positive danger, overhung our father. He was very fearful of going out alone, and he always employed two prize-fighters to act as porters at Pondicherry Lodge. Williams, who drove you tonight, was one of them. He was once light-weight champion of England. Our father would never tell us what it was he feared, but he had a most marked aversion to men with wooden legs. On one occasion he actually fired his revolver at a wooden-legged man, who proved to be a harmless tradesman canvassing for orders. We had to pay a large sum to hush the matter up. My brother and I used to think this a mere whim of my father's, but events have since led us to change our opinion.

"Early in 1882 my father received a letter from India which was a great shock to him. He nearly fainted at the breakfast-table when he opened it, and from that day he sickened to his death. What was in the letter we could never discover, but I could see as he held it that it was short and written in a scrawling hand. He had suffered for years from an enlarged spleen, but he now became rapidly worse, and towards the end of April we were informed that he was beyond all hope, and that he wished to make a last communication to us.

"When we entered his room he was propped up with pillows and breathing heavily. He besought us to lock the door and to come upon either side of the bed. Then, grasping our hands, he made a remarkable statement to us in a voice which was broken as much by emotion as by pain. I shall try and give it to you in his own very words.

"'I have only one thing,' he said, 'which weighs upon my mind at this supreme moment. It is my treatment of poor Morstan's orphan. The cursed greed which has been my besetting sin through life has withheld from her the treasure, half at least of which should have been hers. And yet I have made no use of it myself, so blind and foolish a thing is avarice. The mere feeling of possession has been so dear to me that I could not bear to share it with another. See that chaplet[1] tipped with pearls beside the

---

1  A wreath or garland for the head, usually decorated with a string of beads or pearls.

quinine-bottle. Even that I could not bear to part with, although I had got it out with the design of sending it to her. You, my sons, will give her a fair share of the Agra treasure. But send her nothing—not even the chaplet—until I am gone. After all, men have been as bad as this and have recovered.

"'I will tell you how Morstan died,' he continued. 'He had suffered for years from a weak heart, but he concealed it from every one. I alone knew it. When in India, he and I, through a remarkable chain of circumstances, came into possession of a considerable treasure. I brought it over to England, and on the night of Morstan's arrival he came straight over here to claim his share. He walked over from the station, and was admitted by my faithful old Lal Chowdar, who is now dead. Morstan and I had a difference of opinion as to the division of the treasure, and we came to heated words. Morstan had sprung out of his chair in a paroxysm of anger, when he suddenly pressed his hand to his side, his face turned a dusky hue, and he fell backward, cutting his head against the corner of the treasure-chest. When I stooped over him I found, to my horror, that he was dead.

"'For a long time I sat half distracted, wondering what I should do. My first impulse was, of course, to call for assistance; but I could not but recognize that there was every chance that I would be accused of his murder. His death at the moment of a quarrel, and the gash in his head, would be black against me. Again, an official inquiry could not be made without bringing out some facts about the treasure, which I was particularly anxious to keep secret. He had told me that no soul upon earth knew where he had gone. There seemed to be no necessity why any soul ever should know.

"'I was still pondering over the matter, when, looking up, I saw my servant, Lal Chowdar, in the doorway. He stole in and bolted the door behind him. "Do not fear, Sahib," he said; "no one need know that you have killed him. Let us hide him away, and who is the wiser?" "I did not kill him," said I. Lal Chowdar shook his head and smiled. "I heard it all, Sahib," said he; "I heard you quarrel, and I heard the blow. But my lips are sealed. All are asleep in the house. Let us put him away together." That was enough to decide me. If my own servant could not believe my innocence, how could I hope to make it good before twelve foolish tradesmen in a jury-box? Lal Chowdar and I disposed of the body that night, and within a few days the London papers were full of the mysterious disappearance of Captain Morstan. You will see from what I say that I can hardly be blamed in the matter. My fault lies in the fact that we concealed not only the body but also the treasure and that I have clung to Morstan's share as well as to my own. I wish you, therefore, to make restitution. Put your ears down to my mouth. The treasure is hidden in—'

"At this instant a horrible change came over his expression; his eyes stared wildly, his jaw dropped, and he yelled in a voice which I can never forget, 'Keep him out! For Christ's sake keep him out!' We both stared round at the window behind us upon which his gaze was fixed. A face was looking in at us out of the darkness. We could see the whitening of the nose where it was pressed against the glass. It was a bearded, hairy face, with wild cruel eyes and an expression of concentrated malevolence. My brother and I rushed towards the window, but the man was gone. When we returned to my father his head had dropped and his pulse had ceased to beat.

"We searched the garden that night, but found no sign of the intruder, save that just under the window a single footmark was visible in the flower-bed. But for that one trace, we might have thought that our imaginations had conjured up that wild, fierce face. We soon, however, had another and a more striking proof that there were secret agencies at work all round us. The window of my father's room was found open in the morning, his cupboards and boxes had been rifled, and upon his chest was fixed a torn piece of paper with the words 'The sign of the four' scrawled across it. What the phrase meant, or who our secret visitor may have been, we never knew. As far as we can judge, none of my father's property had been actually stolen, though everything had been turned out. My brother and I naturally associated this peculiar incident with the fear which haunted my father during his life, but it is still a complete mystery to us."

The little man stopped to relight his hookah and puffed thoughtfully for a few moments. We had all sat absorbed, listening to his extraordinary narrative. At the short account of her father's death Miss Morstan had turned deadly white, and for a moment I feared that she was about to faint. She rallied, however, on drinking a glass of water which I quietly poured out for her from a Venetian carafe[1] upon the side-table. Sherlock Holmes leaned back in his chair with an abstracted expression and the lids drawn low over his glittering eyes. As I glanced at him I could

1 Venice is famous for its hand-blown glass; this is yet more evidence of Thaddeus Sholto's conspicuous and extravagant consumption.

not but think how on that very day he had complained bitterly of the commonplaceness of life. Here at least was a problem which would tax his sagacity to the utmost. Mr. Thaddeus Sholto looked from one to the other of us with an obvious pride at the effect which his story had produced and then continued between the puffs of his overgrown pipe.

"My brother and I," said he, "were, as you may imagine, much excited as to the treasure which my father had spoken of. For weeks and for months we dug and delved in every part of the garden without discovering its whereabouts. It was maddening to think that the hiding-place was on his very lips at the moment that he died. We could judge the splendour of the missing riches by the chaplet which he had taken out. Over this chaplet my brother Bartholomew and I had some little discussion. The pearls were evidently of great value, and he was averse to part with them, for, between friends, my brother was himself a little inclined to my father's fault. He thought, too, that if we parted with the chaplet it might give rise to gossip and finally bring us into trouble. It was all that I could do to persuade him to let me find out Miss Morstan's address and send her a detached pearl at fixed intervals so that at least she might never feel destitute."

"It was a kindly thought," said our companion earnestly; "it was extremely good of you."

The little man waved his hand deprecatingly.

"We were your trustees," he said; "that was the view which I took of it, though Brother Bartholomew could not altogether see it in that light. We had plenty of money ourselves. I desired no more. Besides, it would have been such bad taste to have treated a young lady in so scurvy a fashion. '*Le mauvais goût mène au crime.*'[1] The French have a very neat way of putting these things. Our difference of opinion on this subject went so far that I thought it best to set up rooms for myself; so I left Pondicherry Lodge, taking the old khitmutgar[2] and Williams with me. Yesterday, however, I learned that an event of extreme importance has

---

1  French, "bad taste leads to crime," an aphorism credited to the novelist Stendhal (Marie Henri Beyle, 1783-1842).

2  Lal Chowdar, Major Sholto's original *khitmutgar*, was already dead (he predeceased his employer), and Lal Rao remains at Pondicherry Lodge, so it is unclear which butler Thaddeus is referring to here; it could well be the nameless butler now at Brixton who greets Holmes and Watson, but he is nowhere described as "old." There are, at an absolute minimum, at least three different Indian butlers in the Sholtos' employ in London.

occurred. The treasure has been discovered. I instantly communicated with Miss Morstan, and it only remains for us to drive out to Norwood and demand our share. I explained my views last night to Brother Bartholomew, so we shall be expected, if not welcome, visitors."

Mr. Thaddeus Sholto ceased and sat twitching on his luxurious settee. We all remained silent, with our thoughts upon the new development which the mysterious business had taken. Holmes was the first to spring to his feet.

"You have done well, sir, from first to last," said he. "It is possible that we may be able to make you some small return by throwing some light upon that which is still dark to you. But, as Miss Morstan remarked just now, it is late, and we had best put the matter through without delay."

Our new acquaintance very deliberately coiled up the tube of his hookah and produced from behind a curtain a very long befrogged[1] topcoat with Astrakhan[2] collar and cuffs. This he buttoned tightly up, in spite of the extreme closeness of the night, and finished his attire by putting on a rabbit-skin cap with hanging lappets which covered the ears, so that no part of him was visible save his mobile and peaky face.

"My health is somewhat fragile," he remarked as he led the way down the passage. "I am compelled to be a valetudinarian."[3]

Our cab was awaiting us outside, and our programme was evidently prearranged, for the driver started off at once at a rapid pace. Thaddeus Sholto talked incessantly in a voice which rose high above the rattle of the wheels.

"Bartholomew is a clever fellow," said he. "How do you think he found out where the treasure was? He had come to the conclusion that it was somewhere indoors; so he worked out all the cubic space of the house and made measurements everywhere, so that not one inch should be unaccounted for. Among other things, he found that the height of the building was seventy-four feet, but on adding together the heights of all the separate rooms and making every allowance for the space between, which he

---

1  Decorated with spindle shaped buttons which pass through loops from the opposite side of the coat.

2  Named after the largely Tatar city of Astrakhan on the Volga River in southern Russia, this is material made from the skins of still-born or infant lambs. Vladimir Ilyich Lenin's father, who was of Kalmyk Mongol descent, was born in Astrakhan.

3  I.e., an invalid.

ascertained by borings, he could not bring the total to more than seventy feet. There were four feet unaccounted for. These could only be at the top of the building. He knocked a hole, therefore, in the lath and plaster ceiling of the highest room, and there, sure enough, he came upon another little garret above it, which had been sealed up and was known to no one. In the centre stood the treasure-chest resting upon two rafters. He lowered it through the hole, and there it lies. He computes the value of the jewels at not less than half a million sterling."

At the mention of this gigantic sum we all stared at one another open-eyed. Miss Morstan, could we secure her rights, would change from a needy governess to the richest heiress in England. Surely it was the place of a loyal friend to rejoice at such news; yet I am ashamed to say that selfishness took me by the soul, and that my heart turned as heavy as lead within me. I stammered out some few halting words of congratulation, and then sat downcast, with my head drooped, deaf to the babble of our new acquaintance. He was clearly a confirmed hypochondriac, and I was dreamily conscious that he was pouring forth interminable trains of symptoms, and imploring information as to the composition and action of innumerable quack nostrums, some of which he bore about in a leather case in his pocket. I trust that he may not remember any of the answers which I gave him that night. Holmes declares that he overheard me caution him against the great danger of taking more than two drops of castor-oil, while I recommended strychnine in large doses as a sedative. However that may be, I was certainly relieved when our cab pulled up with a jerk and the coachman sprang down to open the door.

"This, Miss Morstan, is Pondicherry Lodge," said Mr. Thaddeus Sholto as he handed her out.

## Chapter 5
### The Tragedy of Pondicherry Lodge

It was nearly eleven o'clock when we reached this final stage of our night's adventures. We had left the damp fog of the great city behind us, and the night was fairly fine. A warm wind blew from the westward, and heavy clouds moved slowly across the sky, with half a moon peeping occasionally through the rifts. It was clear enough to see for some distance, but Thaddeus Sholto took down one of the sidelamps from the carriage to give us a better light upon our way.

Pondicherry Lodge stood in its own grounds and was girt round with a very high stone wall topped with broken glass. A single narrow iron-clamped door formed the only means of entrance. On this our guide knocked with a peculiar postman-like rat-tat.

"Who is there?" cried a gruff voice from within.

"It is I, McMurdo. You surely know my knock by this time."

There was a grumbling sound and a clanking and jarring of keys. The door swung heavily back, and a short, deep-chested man stood in the opening, with the yellow light of the lantern shining upon his protruded face and twinkling, distrustful eyes.

"That you, Mr. Thaddeus? But who are the others? I had no orders about them from the master."

"No, McMurdo? You surprise me! I told my brother last night that I should bring some friends."

"He hain't been out o' his rooms to-day, Mr. Thaddeus, and I have no orders. You know very well that I must stick to regulations. I can let you in, but your friends they must just stop where they are."

This was an unexpected obstacle. Thaddeus Sholto looked about him in a perplexed and helpless manner.

"This is too bad of you, McMurdo!" he said. "If I guarantee them, that is enough for you. There is the young lady, too. She cannot wait on the public road at this hour."

"Very sorry, Mr. Thaddeus," said the porter inexorably. "Folk may be friends o' yours, and yet no friend o' the master's. He pays me well to do my duty, and my duty I'll do. I don't know none o' your friends."

"Oh, yes you do, McMurdo," cried Sherlock Holmes genially. "I don't think you can have forgotten me. Don't you remember that amateur who fought three rounds with you at Alison's rooms on the night of your benefit four years back?"

"Not Mr. Sherlock Holmes!" roared the prize-fighter. "God's truth! how could I have mistook you? If instead o' standin' there so quiet you had just stepped up and given me that cross-hit of yours under the jaw, I'd ha' known you without a question. Ah, you're one that has wasted your gifts, you have! You might have aimed high, if you had joined the fancy."

"You see, Watson, if all else fails me, I have still one of the scientific professions open to me," said Holmes, laughing. "Our friend won't keep us out in the cold now, I am sure."

"In you come, sir, in you come—you and your friends," he

answered. "Very sorry, Mr. Thaddeus, but orders are very strict. Had to be certain of your friends before I let them in."

Inside, a gravel path wound through desolate grounds to a huge clump of a house, square and prosaic, all plunged in shadow save where a moonbeam struck one corner and glimmered in a garret window. The vast size of the building, with its gloom and its deathly silence, struck a chill to the heart. Even Thaddeus Sholto seemed ill at ease, and the lantern quivered and rattled in his hand.

"I cannot understand it," he said. "There must be some mistake. I distinctly told Bartholomew that we should be here, and yet there is no light in his window. I do not know what to make of it."

"Does he always guard the premises in this way?" asked Holmes.

"Yes; he has followed my father's custom. He was the favourite son you know, and I sometimes think that my father may have told him more than he ever told me. That is Bartholomew's window up there where the moonshine strikes. It is quite bright, but there is no light from within, I think."

"None," said Holmes. "But I see the glint of a light in that little window beside the door."

"Ah, that is the housekeeper's room. That is where old Mrs. Bernstone sits. She can tell us all about it. But perhaps you would not mind waiting here for a minute or two, for if we all go in together, and she has had no word of our coming, she may be alarmed. But, hush! what is that?"

He held up the lantern, and his hand shook until the circles of light flickered and wavered all round us. Miss Morstan seized my wrist, and we all stood, with thumping hearts, straining our ears. From the great black house there sounded through the silent night the saddest and most pitiful of sounds—the shrill, broken whimpering of a frightened woman.

"It is Mrs. Bernstone," said Sholto. "She is the only woman in the house. Wait here. I shall be back in a moment."

He hurried for the door and knocked in his peculiar way. We could see a tall old woman admit him and sway with pleasure at the very sight of him.

"Oh, Mr. Thaddeus, sir, I am so glad you have come! I am so glad you have come, Mr. Thaddeus, sir!"

We heard her reiterated rejoicings until the door was closed and her voice died away into a muffled monotone.

Our guide had left us the lantern. Holmes swung it slowly

round and peered keenly at the house and at the great rubbish-heaps which cumbered the grounds. Miss Morstan and I stood together, and her hand was in mine. A wondrous subtle thing is love, for here were we two, who had never seen each other before that day, between whom no word or even look of affection had ever passed, and yet now in an hour of trouble our hands instinctively sought for each other. I have marvelled at it since, but at the time it seemed the most natural thing that I should go out to her so, and, as she has often told me, there was in her also the instinct to turn to me for comfort and protection. So we stood hand in hand like two children, and there was peace in our hearts for all the dark things that surrounded us.

"What a strange place!" she said, looking round.

"It looks as though all the moles in England had been let loose in it. I have seen something of the sort on the side of a hill near Ballarat,[1] where the prospectors had been at work."

"And from the same cause," said Holmes. "These are the traces of the treasure-seekers. You must remember that they were six years looking for it. No wonder that the grounds look like a gravel-pit."

At that moment the door of the house burst open, and Thaddeus Sholto came running out, with his hands thrown forward and terror in his eyes.

"There is something amiss with Bartholomew!" he cried. "I am frightened! My nerves cannot stand it."

He was, indeed, half blubbering with fear, and his twitching, feeble face peeping out from the great Astrakhan collar had the helpless, appealing expression of a terrified child.

"Come into the house," said Holmes in his crisp, firm way.

"Yes, do!" pleaded Thaddeus Sholto, "I really do not feel equal to giving directions."

We all followed him into the housekeeper's room, which stood upon the left-hand side of the passage. The old woman was pacing up and down with a scared look and restless, picking fingers, but the sight of Miss Morstan appeared to have a soothing effect upon her.

"God bless your sweet, calm face!" she cried with a hysterical sob. "It does me good to see you. Oh, but I have been sorely tried this day!"

---

1 Gold mining town in central Australia on the banks of the Yarrowee River, Ballarat was established in 1853 after rich seams of gold were found in the area.

Our companion patted her thin, work-worn hand and murmured some few words of kindly, womanly comfort which brought the colour back into the other's bloodless cheeks.

"Master has locked himself in, and will not answer me," she explained. "All day I have waited to hear from him, for he often likes to be alone; but an hour ago I feared that something was amiss, so I went up and peeped through the keyhole. You must go up, Mr. Thaddeus—you must go up and look for yourself. I have seen Mr. Bartholomew Sholto in joy and in sorrow for ten long years, but I never saw him with such a face on him as that."

Sherlock Holmes took the lamp and led the way, for Thaddeus Sholto's teeth were chattering in his head. So shaken was he that I had to pass my hand under his arm as we went up the stairs, for his knees were trembling under him. Twice as we ascended, Holmes whipped his lens out of his pocket and carefully examined marks which appeared to me to be mere shapeless smudges of dust upon the cocoanut-matting which served as a stair-carpet. He walked slowly from step to step, holding the lamp low, and shooting keen glances to right and left. Miss Morstan had remained behind with the frightened housekeeper.

The third flight of stairs ended in a straight passage of some length, with a great picture in Indian tapestry upon the right of it and three doors upon the left. Holmes advanced along it in the same slow and methodical way, while we kept close at his heels, with our long black shadows streaming backward down the corridor. The third door was that which we were seeking. Holmes knocked without receiving any answer, and then tried to turn the handle and force it open. It was locked on the inside, however, and by a broad and powerful bolt, as we could see when we set our lamp up against it. The key being turned, however, the hole was not entirely closed. Sherlock Holmes bent down to it, and instantly rose again with a sharp intaking of the breath.

"There is something devilish in this, Watson," said he, more moved than I had ever before seen him. "What do you make of it?"

I stooped to the hole and recoiled in horror. Moonlight was streaming into the room, and it was bright with a vague and shifty radiance. Looking straight at me and suspended, as it were, in the air, for all beneath was in shadow, there hung a face—the very face of our companion Thaddeus. There was the same high, shining head, the same circular bristle of red hair, the same bloodless countenance. The features were set, however, in a horrible smile, a fixed and unnatural grin, which in that still and moonlit room was more jarring to the nerves than any scowl or contortion. So like was the face to that of our little friend that I looked round at him to make sure that he was indeed with us. Then I recalled to mind that he had mentioned to us that his brother and he were twins.

"This is terrible!" I said to Holmes. "What is to be done?"

"The door must come down," he answered, and springing against it, he put all his weight upon the lock.

It creaked and groaned, but did not yield. Together we flung ourselves upon it once more, and this time it gave way with a sudden snap, and we found ourselves within Bartholomew Sholto's chamber.

It appeared to have been fitted up as a chemical laboratory. A double line of glass-stoppered bottles was drawn up upon the wall opposite the door, and the table was littered over with Bunsen burners, test-tubes, and retorts. In the corners stood carboys[1] of acid in wicker baskets. One of these appeared to leak or to have been broken, for a stream of dark-coloured liquid had trickled out from it, and the air was heavy with a peculiarly pungent, tar-like odour. A set of steps stood at one side of the room in the midst of a litter of lath and plaster, and above them there was an opening in the ceiling large enough for a man to pass through. At the foot of the steps a long coil of rope was thrown carelessly together.

By the table in a wooden armchair the master of the house was seated all in a heap, with his head sunk upon his left shoulder and that ghastly, inscrutable smile upon his face. He was stiff and cold, and had clearly been dead many hours. It seemed to me that not only his features, but all his limbs, were twisted and turned in the most fantastic fashion. By his hand upon the table there lay a peculiar instrument—a brown, close-grained stick, with a stone head like a hammer, rudely lashed on with coarse twine. Beside it was a torn sheet of note-paper with some words scrawled upon it. Holmes glanced at it and then handed it to me.

"You see," he said with a significant raising of the eyebrows.

In the light of the lantern I read with a thrill of horror, "The sign of the four."

"In God's name, what does it all mean?" I asked.

"It means murder," said he, stooping over the dead man. "Ah! I expected it. Look here!"

---

1  Large globular bottles covered in wicker work.

He pointed to what looked like a long dark thorn stuck in the skin just above the ear.

"It looks like a thorn," said I.

"It is a thorn. You may pick it out. But be careful, for it is poisoned."

I took it up between my finger and thumb. It came away from the skin so readily that hardly any mark was left behind. One tiny speck of blood showed where the puncture had been.

"This is all an insoluble mystery to me," said I. "It grows darker instead of clearer."

"On the contrary," he answered, "it clears every instant. I only require a few missing links to have an entirely connected case."

We had almost forgotten our companion's presence since we entered the chamber. He was still standing in the doorway, the very picture of terror, wringing his hands and moaning to himself. Suddenly, however, he broke out into a sharp, querulous cry.

"The treasure is gone!" he said. "They have robbed him of the treasure! There is the hole through which we lowered it. I helped him to do it! I was the last person who saw him! I left him here last night, and I heard him lock the door as I came downstairs."

"What time was that?"

"It was ten o'clock. And now he is dead, and the police will be called in, and I shall be suspected of having had a hand in it. Oh, yes, I am sure I shall. But you don't think so, gentlemen? Surely you don't think that it was I? Is it likely that I would have brought you here if it were I? Oh, dear! oh, dear! I know that I shall go mad!"

He jerked his arms and stamped his feet in a kind of convulsive frenzy.

"You have no reason for fear, Mr. Sholto," said Holmes kindly, putting his hand upon his shoulder; "take my advice and drive down to the station to report the matter to the police. Offer to assist them in every way. We shall wait here until your return."

The little man obeyed in a half-stupefied fashion, and we heard him stumbling down the stairs in the dark.

## Chapter 6
### Sherlock Holmes Gives a Demonstration

"Now, Watson," said Holmes, rubbing his hands, "we have half an hour to ourselves. Let us make good use of it. My case is, as I have told you, almost complete; but we must not err on the side of over-confidence. Simple as the case seems now, there may be something deeper underlying it."

"Simple!" I ejaculated.

"Surely," said he with something of the air of a clinical professor expounding to his class. "Just sit in the corner there, that your footprints may not complicate matters. Now to work! In the first place, how did these folk come and how did they go? The door has not been opened since last night. How of the window?" He carried the lamp across to it, muttering his observations aloud the while, but addressing them to himself rather than to me. "Window is snibbed[1] on the inner side. Frame-work is solid. No hinges at the side. Let us open it. No water-pipe near. Roof quite out of reach. Yet a man has mounted by the window. It rained a little last night. Here is the print of a foot in mould upon the sill. And here is a circular muddy mark, and here again upon the floor, and here again by the table. See here, Watson! This is really a very pretty demonstration."

I looked at the round, well-defined muddy discs.

"That is not a footmark," said I.

"It is something much more valuable to us. It is the impression of a wooden stump. You see here on the sill is the boot-mark, a heavy boot with a broad metal heel, and beside it is the mark of the timber-toe."

"It is the wooden-legged man."

"Quite so. But there has been someone else—a very able and efficient ally. Could you scale that wall, doctor?"

I looked out of the open window. The moon still shone brightly on that angle of the house. We were a good sixty feet from the ground, and, look where I would, I could see no foothold, nor as much as a crevice in the brickwork.

"It is absolutely impossible," I answered.

"Without aid it is so. But suppose you had a friend up here who lowered you this good stout rope which I see in the corner, securing one end of it to this great hook in the wall. Then, I think, if you were an active man, you might swarm up, wooden leg and all. You would depart, of course, in the same fashion, and your ally would draw up the rope, untie it from the hook, shut the window, snib it on the inside, and get away in the way that he originally came. As a minor point, it may be noted," he continued, fingering the rope, "that our wooden-legged friend, though a fair climber, was not a professional sailor. His hands were far

---

1 Fastened.

from horny. My lens discloses more than one blood-mark, especially towards the end of the rope, from which I gather that he slipped down with such velocity that he took the skin off his hand."

"This is all very well," said I; "but the thing becomes more unintelligible than ever. How about this mysterious ally? How came he into the room?"

"Yes, the ally!" repeated Holmes pensively. "There are features of interest about this ally. He lifts the case from the regions of the commonplace. I fancy that this ally breaks fresh ground in the annals of crime in this country—though parallel cases suggest themselves from India and, if my memory serves me, from Senegambia."[1]

"How came he, then?" I reiterated. "The door is locked; the window is inaccessible. Was it through the chimney?"

"The grate is much too small," he answered. "I had already considered that possibility."

"How, then?" I persisted.

"You will not apply my precept," he said, shaking his head. "How often have I said to you that when you have eliminated the impossible, whatever remains, *however improbable*, must be the truth? We know that he did not come through the door, the window, or the chimney. We also know that he could not have been concealed in the room, as there is no concealment possible. Whence, then, did he come?"

"He came through the hole in the roof!" I cried.

"Of course he did. He must have done so. If you will have the kindness to hold the lamp for me, we shall now extend our researches to the room above—the secret room in which the treasure was found."

He mounted the steps, and, seizing a rafter with either hand, he swung himself up into the garret. Then, lying on his face, he reached down for the lamp and held it while I followed him.

The chamber in which we found ourselves was about ten feet one way and six the other. The floor was formed by the rafters, with thin lath and plaster between, so that in walking one had to step from beam to beam. The roof ran up to an apex, and was evi-

---

1  Region (today a confederation) in West Africa comprising the (at the time) French territory of Senegal and the British territory of the Gambia. The Gambia consists of a strip of land on either side of the Gambia River, and apart from a short coastline on the Atlantic, it is entirely surrounded by Senegal.

dently the inner shell of the true roof of the house. There was no furniture of any sort, and the accumulated dust of years lay thick upon the floor.

"Here you are, you see," said Sherlock Holmes, putting his hand against the sloping wall. "This is a trapdoor which leads out on to the roof. I can press it back, and here is the roof itself, sloping at a gentle angle. This, then, is the way by which Number One entered. Let us see if we can find some other traces of his individuality?"

He held down the lamp to the floor, and as he did so I saw for the second time that night a startled, surprised look come over his face. For myself, as I followed his gaze, my skin was cold under my clothes. The floor was covered thickly with the prints of a naked foot—clear, well-defined, perfectly formed, but scarce half the size of those of an ordinary man.

"Holmes," I said in a whisper, "a child has done this horrid thing."

He had recovered his self-possession in an instant.

"I was staggered for the moment," he said, "but the thing is quite natural. My memory failed me, or I should have been able to foretell it. There is nothing more to be learned here. Let us go down."

"What is your theory, then, as to those footmarks?" I asked eagerly, when we had regained the lower room once more.

"My dear Watson, try a little analysis yourself," said he with a touch of impatience. "You know my methods. Apply them, and it will be instructive to compare results."

"I cannot conceive anything which will cover the facts," I answered.

"It will be clear enough to you soon," he said, in an offhand way. "I think that there is nothing else of importance here, but I will look."

He whipped out his lens and a tape measure, and hurried about the room on his knees, measuring, comparing, examining, with his long thin nose only a few inches from the planks, and his beady eyes gleaming and deep-set like those of a bird. So swift, silent, and furtive were his movements, like those of a trained bloodhound picking out a scent, that I could not but think what a terrible criminal he would have made had he turned his energy and sagacity against the law, instead of exerting them in its defence. As he hunted about, he kept muttering to himself, and finally he broke out into a loud crow of delight.

"We are certainly in luck," said he. "We ought to have very little trouble now. Number One has had the misfortune to tread in the creosote. You can see the outline of the edge of his small foot here at the side of this evil-smelling mess. The carboy has been cracked, you see, and the stuff has leaked out."

"What then?" I asked.

"Why, we have got him, that's all," said he.

"I know a dog that would follow that scent to the world's end. If a pack can track a trailed herring across a shire, how far can a specially trained hound follow so pungent a smell as this? It sounds like a sum in the rule of three. The answer should give us the—But hallo! here are the accredited representatives of the law."

Heavy steps and the clamour of loud voices were audible from below, and the hall door shut with a loud crash.

"Before they come," said Holmes, "just put your hand here on this poor fellow's arm, and here on his leg. What do you feel?"

"The muscles are as hard as a board," I answered.

"Quite so. They are in a state of extreme contraction, far exceeding the usual *rigor mortis*. Coupled with this distortion of the face, this Hippocratic smile, or '*risus sardonicus*,' as the old writers called it, what conclusion would it suggest to your mind?"

"Death from some powerful vegetable alkaloid," I answered, "some strychnine-like substance which would produce tetanus."

"That was the idea which occurred to me the instant I saw the drawn muscles of the face. On getting into the room I at once looked for the means by which the poison had entered the system. As you saw, I discovered a thorn which had been driven or shot with no great force into the scalp. You observe that the part struck was that which would be turned towards the hole in the ceiling if the man were erect in his chair. Now examine this thorn."

I took it up gingerly and held it in the light of the lantern. It was long, sharp, and black, with a glazed look near the point as though some gummy substance had dried upon it. The blunt end had been trimmed and rounded off with a knife.

"Is that an English thorn?" he asked.

"No, it certainly is not."

"With all these data you should be able to draw some just inference. But here are the regulars: so the auxiliary forces may beat a retreat."

As he spoke, the steps which had been coming nearer sounded loudly on the passage, and a very stout, portly man in a gray suit strode heavily into the room. He was red-faced, burly, and plethoric,[1] with a pair of very small twinkling eyes which looked keenly out from between swollen and puffy pouches. He was closely followed by an inspector in uniform, and by the still palpitating Thaddeus Sholto.

"Here's a business!" he cried in a muffled, husky voice. "Here's a pretty business! But who are all these? Why, the house seems to be as full as a rabbit-warren!"

"I think you must recollect me, Mr. Athelney Jones," said Holmes quietly.

"Why, of course I do!" he wheezed. "It's Mr. Sherlock Holmes, the theorist. Remember you! I'll never forget how you lectured us all on causes and inferences and effects in the Bishopgate jewel case. It's true you set us on the right track; but you'll own now that it was more by good luck than good guidance."

"It was a piece of very simple reasoning."

"Oh, come, now, come! Never be ashamed to own up. But what is all this? Bad business! Bad business! Stern facts here—no room for theories. How lucky that I happened to be out at Norwood over another case! I was at the station when the message arrived. What d'you think the man died of?"

"Oh, this is hardly a case for me to theorize over," said Holmes dryly.

"No, no. Still, we can't deny that you hit the nail on the head sometimes. Dear me! Door locked, I understand. Jewels worth half a million missing. How was the window?"

"Fastened; but there are steps on the sill."

"Well, well, if it was fastened the steps could have nothing to do with the matter. That's common sense. Man might have died in a fit; but then the jewels are missing. Ha! I have a theory. These flashes come upon me at times.—Just step outside, Sergeant, and you, Mr. Sholto. Your friend can remain.—What do you think of this, Holmes? Sholto was, on his own confession, with his brother last night. The brother died in a fit, on which Sholto walked off with the treasure? How's that?"

"On which the dead man very considerately got up and locked the door on the inside."

"Hum! There's a flaw there. Let us apply common sense to the matter. This Thaddeus Sholto *was* with his brother; there *was* a quarrel: so much we know. The brother is dead and the jewels are gone. So much also we know. No one saw the brother from

---

1  Ruddy-complexioned; over-fleshy and congested.

the time Thaddeus left him. His bed had not been slept in. Thaddeus is evidently in a most disturbed state of mind. His appearance is—well, not attractive. You see that I am weaving my web round Thaddeus. The net begins to close upon him."

"You are not quite in possession of the facts yet," said Holmes. "This splinter of wood, which I have every reason to believe to be poisoned, was in the man's scalp where you still see the mark; this card, inscribed as you see it, was on the table, and beside it lay this rather curious stone-headed instrument. How does all that fit into your theory?"

"Confirms it in every respect," said the fat detective pompously. "House is full of Indian curiosities. Thaddeus brought this up, and if this splinter be poisonous Thaddeus may as well have made murderous use of it as any other man. The card is some hocus-pocus—a blind, as like as not. The only question is, how did he depart? Ah, of course, here is a hole in the roof."

With great activity, considering his bulk, he sprang up the steps and squeezed through into the garret, and immediately afterwards we heard his exulting voice proclaiming that he had found the trap-door.

"He can find something," remarked Holmes, shrugging his shoulders; "he has occasional glimmerings of reason. *Il n'y a pas des sots si incommodes que ceux qui ont de l'esprit!* "[1]

"You see!" said Athelney Jones, reappearing down the steps again; "facts are better than theories, after all. My view of the case is confirmed. There is a trapdoor communicating with the roof, and it is partly open."

"It was I who opened it."

"Oh, indeed! You did notice it, then?" He seemed a little crestfallen at the discovery. "Well, whoever noticed it, it shows how our gentleman got away. Inspector!"

"Yes, sir," from the passage.

"Ask Mr. Sholto to step this way.—Mr. Sholto, it is my duty to inform you that anything which you may say will be used against you. I arrest you in the Queen's name as being concerned in the death of your brother."

"There, now! Didn't I tell you!" cried the poor little man throwing out his hands and looking from one to the other of us.

---

1 French, "there are no fools so troublesome as those who have wit"; from *The Maxims and Reflections* of François, Duc de la Rochefoucauld (1613-80).

"Don't trouble yourself about it, Mr. Sholto," said Holmes; "I think that I can engage to clear you of the charge."

"Don't promise too much, Mr. Theorist, don't promise too much!" snapped the detective. "You may find it a harder matter than you think."

"Not only will I clear him, Mr. Jones, but I will make you a free present of the name and description of one of the two people who were in this room last night. His name, I have every reason to believe, is Jonathan Small. He is a poorly-educated man, small, active, with his right leg off, and wearing a wooden stump which is worn away upon the inner side. His left boot has a coarse, square-toed sole, with an iron band round the heel. He is a middle-aged man, much sunburned, and has been a convict. These few indications may be of some assistance to you, coupled with the fact that there is a good deal of skin missing from the palm of his hand. The other man—"

"Ah! the other man?" asked Athelney Jones in a sneering voice, but impressed none the less, as I could easily see, by the precision of the other's manner.

"Is a rather curious person," said Sherlock Holmes, turning upon his heel. "I hope before very long to be able to introduce you to the pair of them. A word with you, Watson."

He led me out to the head of the stair.

"This unexpected occurrence," he said, "has caused us rather to lose sight of the original purpose of our journey."

"I have just been thinking so," I answered; "it is not right that Miss Morstan should remain in this stricken house."

"No. You must escort her home. She lives with Mrs. Cecil Forrester, in Lower Camberwell,[1] so it is not very far. I will wait for you here if you will drive out again. Or perhaps you are too tired?"

"By no means. I don't think I could rest until I know more of this fantastic business. I have seen something of the rough side of life, but I give you my word that this quick succession of strange surprises to-night has shaken my nerve completely. I should like, however, to see the matter through with you, now that I have got so far."

"Your presence will be of great service to me," he answered. "We shall work the case out independently, and leave this fellow

---

1 Another, slightly less respectable, middle-class South London suburb, made famous by George Gissing in his novel, *In the Year of Jubilee* (1894).

Jones to exult over any mare's-nest which he may choose to construct. When you have dropped Miss Morstan, I wish you to go on to No. 3, Pinchin Lane, down near the water's edge at Lambeth.[1] The third house on the right-hand side is a bird-stuffer's; Sherman is the name. You will see a weasel holding a young rabbit in the window. Knock old Sherman up and tell him, with my compliments, that I want Toby at once. You will bring Toby back in the cab with you."

"A dog, I suppose."

"Yes, a queer mongrel, with a most amazing power of scent. I would rather have Toby's help than that of the whole detective force of London."

"I shall bring him then," said I. "It is one now. I ought to be back before three, if I can get a fresh horse."

"And I," said Holmes, "shall see what I can learn from Mrs. Bernstone, and from the Indian servant, who, Mr. Thaddeus tells me, sleeps in the next garret. Then I shall study the great Jones's methods and listen to his not too delicate sarcasms. '*Wir sind gewohnt dass die Menschen verhöhnen was sie nicht verstehen.*'[2] Goethe is always pithy."

## Chapter 7
### The Episode of the Barrel

The police had brought a cab with them, and in this I escorted Miss Morstan back to her home. After the angelic fashion of women, she had borne trouble with a calm face as long as there was someone weaker than herself to support, and I had found her bright and placid by the side of the frightened housekeeper. In the cab, however, she first turned faint and then burst into a passion of weeping—so sorely had she been tried by the adventures of the night. She has told me since that she thought me cold and distant upon that journey. She little guessed the struggle within my breast, or the effort of self-restraint which held me back. My sympathies and my love went out to her, even as my

---

1 Another South London district, this time on the south bank of the Thames, immediately opposite the Palace of Westminster. London's Waterloo Station and Lambeth Palace, the Archbishop of Canterbury's official residence, are both in the area.

2 German, "We are used to seeing that man despises what he never comprehends"; from *Faust*, Part I, by Johann Wolfgang von Goethe (1749-1832).

hand had in the garden. I felt that years of the conventionalities of life could not teach me to know her sweet, brave nature as had this one day of strange experiences. Yet there were two thoughts which sealed the words of affection upon my lips. She was weak and helpless, shaken in mind and nerve. It was to take her at a disadvantage to obtrude love upon her at such a time. Worse still, she was rich. If Holmes's researches were successful, she would be an heiress. Was it fair, was it honourable, that a half-pay surgeon should take such advantage of an intimacy which chance had brought about? Might she not look upon me as a mere vulgar fortune-seeker? I could not bear to risk that such a thought should cross her mind. This Agra treasure intervened like an impassable barrier between us.

It was nearly two o'clock when we reached Mrs. Cecil Forrester's. The servants had retired hours ago, but Mrs. Forrester had been so interested by the strange message which Miss Morstan had received that she had sat up in the hope of her return. She opened the door herself, a middle-aged, graceful woman, and it gave me joy to see how tenderly her arm stole round the other's waist and how motherly was the voice in which she greeted her. She was clearly no mere paid dependant, but an honoured friend. I was introduced, and Mrs. Forrester earnestly begged me to step in and tell her our adventures. I explained, however, the importance of my errand and promised faithfully to call and report any progress which we might make with the case. As we drove away I stole a glance back, and I still seem to see that little group on the step—the two graceful, clinging figures, the half-opened door, the hall-light shining through stained glass, the barometer, and the bright stair-rods. It was soothing to catch even that passing glimpse of a tranquil English home in the midst of the wild, dark business which had absorbed us.

And the more I thought of what had happened, the wilder and darker it grew. I reviewed the whole extraordinary sequence of events as I rattled on through the silent, gas-lit streets. There was the original problem: that at least was pretty clear now. The death of Captain Morstan, the sending of the pearls, the advertisement, the letter—we had had light upon all those events. They had only led us, however, to a deeper and far more tragic mystery. The Indian treasure, the curious plan found among Morstan's baggage, the strange scene at Major Sholto's death, the rediscovery of the treasure immediately followed by the murder of the discoverer, the very singular accompaniments to the crime, the footsteps, the remarkable weapons, the words upon the card, corresponding with those upon Captain Morstan's chart—here was indeed a labyrinth in which a man less singularly endowed than my fellow-lodger might well despair of ever finding the clue.

Pinchin Lane was a row of shabby, two-storied brick houses in the lower quarter of Lambeth. I had to knock for some time at No. 3 before I could make any impression. At last, however, there was the glint of a candle behind the blind, and a face looked out at the upper window.

"Go on, you drunken vagabond," said the face. "If you kick up any more row, I'll open the kennels and let out forty-three dogs upon you."

"If you'll let one out, it's just what I have come for," said I.

"Go on!" yelled the voice. "So help me gracious, I have a wiper[1] in this bag, and I'll drop it on your 'ead if you don't hook it!"

"But I want a dog," I cried.

"I won't be argued with!" shouted Mr. Sherman. "Now stand clear, for when I say 'three,' down goes the wiper."

"Mr. Sherlock Holmes—" I began; but the words had a most magical effect, for the window instantly slammed down, and within a minute the door was unbarred and open. Mr. Sherman was a lanky, lean old man, with stooping shoulders, a stringy neck, and blue-tinted glasses.

"A friend of Mr. Sherlock is always welcome," said he. "Step in, sir. Keep clear of the badger, for he bites. Ah, naughty, naughty! would you take a nip at the gentleman?" This to a stoat which thrust its wicked head and red eyes between the bars of its cage. "Don't mind that, sir; it's only a slowworm. It hain't got no fangs, so I gives it the run o' the room, for it keeps the beetles down. You must not mind my bein' just a little short wi' you at first, for I'm guyed at by the children, and there's many a one just comes down this lane to knock me up. What was it that Mr. Sherlock Holmes wanted, sir?"

"He wanted a dog of yours."

"Ah! that would be Toby."

"Yes, Toby was the name."

"Toby lives at No. 7 on the left here."

He moved slowly forward with his candle among the queer animal family which he had gathered round him. In the uncertain, shadowy light I could see dimly that there were glancing, glimmering eyes peeping down at us from every cranny and

---

1   A viper, in Sherman's thick cockney accent.

corner. Even the rafters above our heads were lined by solemn fowls, who lazily shifted their weight from one leg to the other as our voices disturbed their slumbers.

Toby proved to be an ugly, long-haired, lop-eared creature, half spaniel and half lurcher, brown and white in colour, with a very clumsy, waddling gait. It accepted, after some hesitation, a lump of sugar which the old naturalist handed to me, and, having thus sealed an alliance, it followed me to the cab, and made no difficulties about accompanying me. It had just struck three on the Palace clock[1] when I found myself back once more at Pondicherry Lodge. The ex-prize-fighter McMurdo had, I found, been arrested as an accessory, and both he and Mr. Sholto had been marched off to the station. Two constables guarded the narrow gate, but they allowed me to pass with the dog on my mentioning the detective's name.

Holmes was standing on the doorstep with his hands in his pockets, smoking his pipe.

"Ah, you have him there!" said he. "Good dog, then! Athelney Jones has gone. We have had an immense display of energy since you left. He has arrested not only friend Thaddeus but the gatekeeper, the housekeeper, and the Indian servant. We have the place to ourselves but for a sergeant upstairs. Leave the dog here and come up."

We tied Toby to the hall table, and reascended the stairs. The room was as we had left it, save that a sheet had been draped over the central figure. A weary-looking police-sergeant reclined in the corner.

"Lend me your bull's eye,[2] Sergeant," said my companion. "Now tie this bit of card round my neck, so as to hang it in front of me. Thank you. Now I must kick off my boots and stockings. Just you carry them down with you, Watson. I am going to do a little climbing. And dip my handkerchief into the creosote. That will do. Now come up into the garret with me for a moment."

We clambered up through the hole. Holmes turned his light once more upon the footsteps in the dust.

"I wish you particularly to notice these footmarks," he said. "Do you observe anything noteworthy about them?"

---

1  The Palace of Westminster's clock, popularly known as Big Ben. Of course, Watson is using this term metaphorically, as he would not have been able to hear the clock strike from Pondicherry Lodge, or even from Lambeth.
2  Lantern.

"They belong," I said, "to a child or a small woman."

"Apart from their size, though. Is there nothing else?"

"They appear to be much as other footmarks."

"Not at all. Look here! This is the print of a right foot in the dust. Now I make one with my naked foot beside it. What is the chief difference?"

"Your toes are all cramped together. The other print has each toe distinctly divided."

"Quite so. That is the point. Bear that in mind. Now, would you kindly step over to that flap-window and smell the edge of the woodwork? I shall stay over here, as I have this handkerchief in my hand."

I did as he directed and was instantly conscious of a strong tarry smell.

"That is where he put his foot in getting out. If *you* can trace him, I should think that Toby will have no difficulty. Now run downstairs, loose the dog, and look out for Blondin."[1]

By the time that I got out into the grounds Sherlock Holmes was on the roof, and I could see him like an enormous glow-worm crawling very slowly along the ridge. I lost sight of him behind a stack of chimneys, but he presently reappeared and then vanished once more upon the opposite side. When I made my way round there I found him seated at one of the corner eaves.

"That you, Watson?" he cried.

"Yes."

"This is the place. What is that black thing down there?"

"A water-barrel."

"Top on it?"

"Yes."

"No sign of a ladder?"

"No."

"Confound the fellow! It's a most breakneck place. I ought to be able to come down where he could climb up. The water-pipe feels pretty firm. Here goes, anyhow."

There was a scuffling of feet, and the lantern began to come steadily down the side of the wall. Then with a light spring he came on to the barrel, and from there to the earth.

"It was easy to follow him," he said, drawing on his stockings and boots. "Tiles were loosened the whole way along, and in his

---

1 Stage name of French acrobat Jean-François Gravelet (1824-97), who had famously crossed the Niagara Falls on a tightrope in 1859.

hurry he had dropped this. It confirms my diagnosis, as you doctors express it."

The object which he held up to me was a small pocket or pouch woven out of coloured grasses and with a few tawdry beads strung round it. In shape and size it was not unlike a cigarette-case. Inside were half a dozen spines of dark wood, sharp at one end and rounded at the other, like that which had struck Bartholomew Sholto.

"They are hellish things," said he. "Look out that you don't prick yourself. I'm delighted to have them, for the chances are that they are all he has. There is the less fear of you or me finding one in our skin before long. I would sooner face a Martini[1] bullet, myself. Are you game for a six-mile trudge, Watson?"

"Certainly," I answered.

"Your leg will stand it?"

"Oh, yes."

"Here you are, doggy! Good old Toby! Smell it, Toby, smell it!" He pushed the creosote handkerchief under the dog's nose, while the creature stood with its fluffy legs separated, and with a most comical cock to its head, like a connoisseur sniffing the *bouquet* of a famous vintage. Holmes then threw the handkerchief to a distance, fastened a stout cord to the mongrel's collar, and led him to the foot of the water-barrel. The creature instantly broke into a succession of high, tremulous yelps and, with his nose on the ground and his tail in the air, pattered off upon the trail at a pace which strained his leash and kept us at the top of our speed.

The east had been gradually whitening, and we could now see some distance in the cold gray light. The square, massive house, with its black, empty windows and high, bare walls, towered up, sad and forlorn, behind us. Our course led right across the grounds, in and out among the trenches and pits with which they were scarred and intersected. The whole place, with its scattered dirt-heaps and ill-grown shrubs, had a blighted, ill-omened look which harmonized with the black tragedy which hung over it.

On reaching the boundary wall Toby ran along, whining eagerly, underneath its shadow, and stopped finally in a corner screened by a young beech. Where the two walls joined, several bricks had been loosened, and the crevices left were worn down and rounded upon the lower side, as though they had frequently

---

1 Single-shot breech mechanism for rifles, invented in the 1860s and used successfully in the Second Anglo-Afghan War, 1878-80.

been used as a ladder. Holmes clambered up, and taking the dog from me, he dropped it over upon the other side.

"There's the print of Wooden-leg's hand," he remarked as I mounted up beside him. "You see the slight smudge of blood upon the white plaster. What a lucky thing it is that we have had no very heavy rain since yesterday! The scent will lie upon the road in spite of their eight-and-twenty hours' start."

I confess that I had my doubts myself when I reflected upon the great traffic which had passed along the London road in the interval. My fears were soon appeased, however. Toby never hesitated or swerved but waddled on in his peculiar rolling fashion. Clearly the pungent smell of the creosote rose high above all other contending scents.

"Do not imagine," said Holmes, "that I depend for my success in this case upon the mere chance of one of these fellows having put his foot in the chemical. I have knowledge now which would enable me to trace them in many different ways. This, however, is the readiest, and, since fortune has put it into our hands, I should be culpable if I neglected it. It has, however prevented the case from becoming the pretty little intellectual problem which it at one time promised to be. There might have been some credit to be gained out of it, but for this too palpable clue."

"There is credit, and to spare," said I. "I assure you, Holmes, that I marvel at the means by which you obtain your results in this case, even more than I did in the Jefferson Hope murder. The thing seems to me to be deeper and more inexplicable. How, for example, could you describe with such confidence the wooden-legged man?"

"Pshaw, my dear boy! it was simplicity itself. I don't wish to be theatrical. It is all patent and above-board. Two officers who are in command of a convict-guard learn an important secret as to buried treasure. A map is drawn for them by an Englishman named Jonathan Small. You remember that we saw the name upon the chart in Captain Morstan's possession. He had signed it on behalf of himself and his associates—the sign of the four, as he somewhat dramatically called it. Aided by this chart, the officers—or one of them—gets the treasure and brings it to England, leaving, we will suppose, some condition under which he received it unfulfilled. Now, then, why did not Jonathan Small get the treasure himself? The answer is obvious. The chart is dated at a time when Morstan was brought into close association with convicts. Jonathan Small did not get the treasure because he and his associates were themselves convicts and could not get away."

"But this is mere speculation," said I.

"It is more than that. It is the only hypothesis which covers the facts. Let us see how it fits in with the sequel. Major Sholto remains at peace for some years, happy in the possession of his treasure. Then he receives a letter from India which gives him a great fright. What was that?"

"A letter to say that the men whom he had wronged had been set free."

"Or had escaped. That is much more likely, for he would have known what their term of imprisonment was. It would not have been a surprise to him. What does he do then? He guards himself against a wooden-legged man—a white man, mark you, for he mistakes a white tradesman for him and actually fires a pistol at him. Now, only one white man's name is on the chart. The others are Hindoos or Mohammedans.[1] There is no other white man. Therefore we may say with confidence that the wooden-legged man is identical with Jonathan Small. Does the reasoning strike you as being faulty?"

"No: it is clear and concise."

"Well, now, let us put ourselves in the place of Jonathan Small. Let us look at it from his point of view. He comes to England with the double idea of regaining what he would consider to be his rights and of having his revenge upon the man who had wronged him. He found out where Sholto lived, and very possibly he established communications with someone inside the house. There is this butler, Lal Rao,[2] whom we have not seen. Mrs. Bernstone gives him far from a good character. Small could not find out, however, where the treasure was hid, for no one ever knew save the major and one faithful servant who had died. Suddenly Small learns that the major is on his deathbed. In a frenzy lest the secret of the treasure die with him, he runs the gauntlet of the guards, makes his way to the dying man's window, and is only deterred from entering by the presence of his two sons. Mad

---

1   Not quite—the names (Dost Akbar, Mahomet Singh, and Abdullah Khan) are Muslim or possibly Sikh (and in Mahomet Singh's case, an unlikely conflation of the two), but certainly not Hindu.

2   This is evidently Bartholomew Sholto's Indian butler at Pondicherry Lodge, and not Thaddeus Sholto's nameless *khitmutgar* in Brixton. Major John Sholto's late butler, who helped him dispose of Morstan's body, was called Lal Chowdar; he died after Morstan's death on 3 December1878 and before Major Sholto's on 28 April 1882. Sholto clearly returned from India with a retinue of servants.

with hate, however, against the dead man, he enters the room that night, searches his private papers in the hope of discovering some memorandum relating to the treasure, and finally leaves a memento of his visit in the short inscription upon the card. He had doubtless planned beforehand that, should he slay the major, he would leave some such record upon the body as a sign that it was not a common murder but, from the point of view of the four associates, something in the nature of an act of justice. Whimsical and bizarre conceits of this kind are common enough in the annals of crime and usually afford valuable indications as to the criminal. Do you follow all this?"

"Very clearly."

"Now what could Jonathan Small do? He could only continue to keep a secret watch upon the efforts made to find the treasure. Possibly he leaves England and only comes back at intervals. Then comes the discovery of the garret, and he is instantly informed of it. We again trace the presence of some confederate in the household. Jonathan, with his wooden leg, is utterly unable to reach the lofty room of Bartholomew Sholto. He takes with him, however, a rather curious associate, who gets over this difficulty but dips his naked foot into creosote, whence come Toby, and a six-mile limp for a half-pay officer with a damaged tendo Achillis."

"But it was the associate and not Jonathan who committed the crime."

"Quite so. And rather to Jonathan's disgust, to judge by the way he stamped about when he got into the room. He bore no grudge against Bartholomew Sholto and would have preferred if he could have been simply bound and gagged. He did not wish to put his head in a halter. There was no help for it, however: the savage instincts of his companion had broken out, and the poison had done its work: so Jonathan Small left his record, lowered the treasure-box to the ground, and followed it himself. That was the train of events as far as I can decipher them. Of course, as to his personal appearance, he must be middle-aged and must be sun-burned after serving his time in such an oven as the Andamans. His height is readily calculated from the length of his stride, and we know that he was bearded. His hairiness was the one point which impressed itself upon Thaddeus Sholto when he saw him at the window. I don't know that there is anything else."

"The associate?"

"Ah, well, there is no great mystery in that. But you will know all about it soon enough. How sweet the morning air is! See how that one little cloud floats like a pink feather from some gigantic flamingo. Now the red rim of the sun pushes itself over the London cloud-bank. It shines on a good many folk, but on none, I dare bet, who are on a stranger errand than you and I. How small we feel with our petty ambitions and strivings in the presence of the great elemental forces of Nature! Are you well up in your Jean Paul?"[1]

"Fairly so. I worked back to him through Carlyle."

"That was like following the brook to the parent lake. He makes one curious but profound remark. It is that the chief proof of man's real greatness lies in his perception of his own smallness. It argues, you see, a power of comparison and of appreciation which is in itself a proof of nobility. There is much food for thought in Richter. You have not a pistol, have you?"

"I have my stick."

"It is just possible that we may need something of the sort if we get to their lair. Jonathan I shall leave to you, but if the other turns nasty I shall shoot him dead."

He took out his revolver as he spoke, and, having loaded two of the chambers, he put it back into the right-hand pocket of his jacket.

We had during this time been following the guidance of Toby down the half-rural villa-lined roads which lead to the metropolis. Now, however, we were beginning to come among continuous streets, where labourers and dockmen were already astir, and slatternly women were taking down shutters and brushing doorsteps. At the square-topped corner public-houses business was just beginning, and rough-looking men were emerging, rubbing their sleeves across their beards after their morning wet. Strange dogs sauntered up and stared wonderingly at us as we passed, but our inimitable Toby looked neither to the right nor to the left but trotted onward with his nose to the ground and an occasional eager whine which spoke of a hot scent.

We had traversed Streatham, Brixton, Camberwell, and now found ourselves in Kennington Lane, having borne away through the side streets to the east of the Oval.[2] The men whom we

1  Pseudonym of German Romantic novelist Johann Paul Friedrich Richter (1763-1825). His work was translated and popularised in the English speaking world by (amongst others) the Scottish historian and essayist, Thomas Carlyle (1795-1881), especially in his *German Romances* (1858).

2  A journey traversing much of South London, in a steadily northward direction, and culminating close to the famous cricket ground, the Oval.

pursued seemed to have taken a curiously zigzag road, with the idea probably of escaping observation. They had never kept to the main road if a parallel side street would serve their turn. At the foot of Kennington Lane they had edged away to the left through Bond Street and Miles Street.[1] Where the latter street turns into Knight's Place, Toby ceased to advance but began to run backward and forward with one ear cocked and the other drooping, the very picture of canine indecision. Then he waddled round in circles, looking up to us from time to time, as if to ask for sympathy in his embarrassment.

"What the deuce is the matter with the dog?" growled Holmes. "They surely would not take a cab or go off in a balloon."

"Perhaps they stood here for some time," I suggested.

"Ah! it's all right. He's off again," said my companion in a tone of relief.

He was indeed off, for after sniffing round again he suddenly made up his mind, and darted away with an energy and determination such as he had not yet shown. The scent appeared to be much hotter than before, for he had not even to put his nose on the ground, but tugged at his leash and tried to break into a run. I could see by the gleam in Holmes's eyes that he thought we were nearing the end of our journey.

Our course now ran down Nine Elms[2] until we came to Broderick and Nelson's large timber-yard, just past the White Eagle tavern. Here the dog, frantic with excitement, turned down through the side gate into the enclosure, where the sawyers were already at work. On the dog raced through sawdust and shavings, down an alley, round a passage, between two wood-piles, and finally, with a triumphant yelp, sprang upon a large barrel which still stood upon the hand-trolley on which it had been brought. With lolling tongue and blinking eyes Toby stood upon the cask, looking from one to the other of us for some sign of appreciation. The staves of the barrel and the wheels of the trolley were smeared with a dark liquid, and the whole air was heavy with the smell of creosote.

Sherlock Holmes and I looked blankly at each other and then burst simultaneously into an uncontrollable fit of laughter.

---

1  Two small streets immediately south of Vauxhall station.
2  Nine Elm's Road, which runs between Vauxhall and Battersea, roughly parallel to the Thames.

## Chapter 8
### The Baker Street Irregulars

"What now?" I asked. "Toby has lost his character for infallibility."

"He acted according to his lights," said Holmes, lifting him down from the barrel and walking him out of the timber-yard. "If you consider how much creosote is carted about London in one day, it is no great wonder that our trail should have been crossed. It is much used now, especially for the seasoning of wood. Poor Toby is not to blame."

"We must get on the main scent again, I suppose."

"Yes. And, fortunately, we have no distance to go. Evidently what puzzled the dog at the corner of Knight's Place was that there were two different trails running in opposite directions. We took the wrong one. It only remains to follow the other."

There was no difficulty about this. On leading Toby to the place where he had committed his fault, he cast about in a wide circle and finally dashed off in a fresh direction.

"We must take care that he does not now bring us to the place where the creosote-barrel came from," I observed.

"I had thought of that. But you notice that he keeps on the pavement, whereas the barrel passed down the roadway. No, we are on the true scent now."

It tended down towards the river-side, running through Belmont Place and Prince's Street. At the end of Broad Street it ran right down to the water's edge, where there was a small wooden wharf. Toby led us to the very edge of this and there stood whining, looking out on the dark current beyond.

"We are out of luck," said Holmes. "They have taken to a boat here."

Several small punts and skiffs were lying about in the water and on the edge of the wharf. We took Toby round to each in turn, but though he sniffed earnestly, he made no sign.

Close to the rude landing-stage was a small brick house, with a wooden placard slung out through the second window. "Mordecai Smith" was printed across it in large letters, and, underneath, "Boats to hire by the hour or day." A second inscription above the door informed us that a steam launch was kept— a statement which was confirmed by a great pile of coke upon the jetty. Sherlock Holmes looked slowly round, and his face assumed an ominous expression.

"This looks bad," said he. "These fellows are sharper than I expected. They seem to have covered their tracks. There has, I fear, been preconcerted management here."

He was approaching the door of the house, when it opened, and a little curly-headed lad of six came running out, followed by a stoutish, red-faced woman with a large sponge in her hand.

"You come back and be washed, Jack," she shouted. "Come back, you young imp; for if your father comes home and finds you like that he'll let us hear of it."

"Dear little chap!" said Holmes strategically. "What a rosy-cheeked young rascal! Now, Jack, is there anything you would like?"

The youth pondered for a moment.

"I'd like a shillin'," said he.

"Nothing you would like better?"

"I'd like two shillin' better," the prodigy answered after some thought.

"Here you are, then! Catch!—A fine child, Mrs. Smith!"

"Lor' bless you, sir, he is that, and forward. He gets a'most too much for me to manage, 'specially when my man is away days at a time."

"Away, is he?" said Holmes in a disappointed voice. "I am sorry for that, for I wanted to speak to Mr. Smith."

"He's been away since yesterday mornin', sir, and, truth to tell, I am beginnin' to feel frightened about him. But if it was about a boat, sir, maybe I could serve as well."

"I wanted to hire his steam launch."

"Why, bless you, sir, it is in the steam launch that he has gone. That's what puzzles me, for I know there ain't more coals in her than would take her to about Woolwich[1] and back. If he's been away in the barge I'd ha' thought nothin'; for many a time a job has taken him as far as Gravesend,[2] and then if there was much doin' there he might ha' stayed over. But what good is a steam launch without coals?"

"He might have bought some at a wharf down the river."

"He might, sir, but it weren't his way. Many a time I've heard him call out at the prices they charge for a few odd bags. Besides,

---

1  Area of southeast London on the south bank of the Thames downstream from Lambeth.
2  A small town in Kent, on the south bank of the Thames estuary. Charles Dickens lived in nearby Gad's Hill Place from 1857 until his death in 1870.

I don't like that wooden-legged man, wi' his ugly face and out-
landish talk. What did he want always knockin' about here for?"

"A wooden-legged man?" said Holmes with bland surprise.

"Yes, sir, a brown, monkey-faced chap that's called more'n
once for my old man. It was him that roused him up yesternight
and, what's more, my man knew he was comin', for he had steam
up in the launch. I tell you straight, sir, I don't feel easy in my
mind about it."

"But, my dear Mrs. Smith," said Holmes, shrugging his
shoulders, "you are frightening yourself about nothing. How
could you possibly tell that it was the wooden-legged man who
came in the night? I don't quite understand how you can be so
sure."

"His voice, sir. I knew his voice, which is kind o' thick and
foggy. He tapped at the winder—about three it would be. 'Show
a leg, matey,' says he: 'time to turn out guard.' My old man woke
up Jim—that's my eldest—and away they went without so much
as a word to me. I could hear the wooden leg clackin' on the
stones."

"And was this wooden-legged man alone?"

"Couldn't say, I am sure, sir. I didn't hear no one else."

"I am sorry, Mrs. Smith, for I wanted a steam launch, and I
have heard good reports of the—Let me see, what is her name?"

"The *Aurora*,[1] sir."

"Ah! She's not that old green launch with a yellow line, very
broad in the beam?"

"No, indeed. She's as trim a little thing as any on the river.
She's been fresh painted, black with two red streaks."

"Thanks. I hope that you will hear soon from Mr. Smith. I
am going down the river, and if I should see anything of the
Aurora I shall let him know that you are uneasy. A black funnel,
you say?"

"No, sir. Black with a white band."

"Ah, of course. It was the sides which were black. Good-
morning, Mrs. Smith. There is a boatman here with a wherry,[2]
Watson. We shall take it and cross the river."

"The main thing with people of that sort," said Holmes as we
sat in the sheets of the wherry, "is never to let them think that
their information can be of the slightest importance to you. If you

---

1 Dawn (Latin), a particularly suitable name, for Small commandeers the
  boat just before dawn.
2 A light rowboat, usually used to carry passengers across rivers.

do they will instantly shut up like an oyster. If you listen to them under protest, as it were, you are very likely to get what you want."

"Our course now seems pretty clear," said I.

"What would you do, then?"

"I would engage a launch and go down the river on the track of the *Aurora*."

"My dear fellow, it would be a colossal task. She may have touched at any wharf on either side of the stream between here and Greenwich.[1] Below the bridge there is a perfect labyrinth of landing-places for miles. It would take you days and days to exhaust them if you set about it alone."

"Employ the police, then."

"No. I shall probably call Athelney Jones in at the last moment. He is not a bad fellow, and I should not like to do anything which would injure him professionally. But I have a fancy for working it out myself, now that we have gone so far."

"Could we advertise, then, asking for information from wharfingers?"

"Worse and worse! Our men would know that the chase was hot at their heels, and they would be off out of the country. As it is, they are likely enough to leave, but as long as they think they are perfectly safe they will be in no hurry. Jones's energy will be of use to us there, for his view of the case is sure to push itself into the daily press, and the runaways will think that everyone is off on the wrong scent."

"What are we to do, then?" I asked as we landed near Millbank Penitentiary.[2]

---

1  Southeast London suburb on the south bank of the Thames, upstream from Woolwich. Home of the Royal Observatory and birthplace of Greenwich Mean Time, it is now inscribed as a UNESCO World Heritage Site.

2  Designed by architect William Williams according to the principles of the utilitarian philosopher Jeremy Bentham and opened in 1821, Millbank Penitentiary, an octagonal shaped panopticon, was London's largest prison. It was used for one purpose only: to house prisoners who had been sentenced to be transported to the colonies for the rest of their natural lives. Every inmate in the 1,100 cells could be viewed from the central Governor's office. Prisoners were kept under close surveillance for three months in order to determine to which colony they would be transported; in 1850, some 4,000 prisoners passed through the facility. Millbank Penitentiary was used until 1886 and demolished in 1890; the Tate Britain Gallery is partly built upon its grounds.

"Take this hansom, drive home, have some breakfast, and get an hour's sleep. It is quite on the cards that we may be afoot to-night again. Stop at a telegraph office, cabby! We will keep Toby, for he may be of use to us yet."

We pulled up at the Great Peter Street[1] Post-Office, and Holmes dispatched his wire.

"Whom do you think that is to?" he asked as we resumed our journey.

"I am sure I don't know."

"You remember the Baker Street division of the detective police force whom I employed in the Jefferson Hope case?"

"Well," said I, laughing.

"This is just the case where they might be invaluable. If they fail I have other resources, but I shall try them first. That wire was to my dirty little lieutenant, Wiggins, and I expect that he and his gang will be with us before we have finished our breakfast."

It was between eight and nine o'clock now, and I was conscious of a strong reaction after the successive excitements of the night. I was limp and weary, befogged in mind and fatigued in body. I had not the professional enthusiasm which carried my companion on, nor could I look at the matter as a mere abstract intellectual problem. As far as the death of Bartholomew Sholto went, I had heard little good of him and could feel no intense antipathy to his murderers. The treasure, however, was a different matter. That, or part of it, belonged rightfully to Miss Morstan. While there was a chance of recovering it I was ready to devote my life to the one object. True, if I found it, it would probably put her forever beyond my reach. Yet it would be a petty and selfish love which would be influenced by such a thought as that. If Holmes could work to find the criminals, I had a tenfold stronger reason to urge me on to find the treasure.

A bath at Baker Street and a complete change freshened me up wonderfully. When I came down to our room I found the breakfast laid and Holmes pouring out the coffee.

"Here it is," said he, laughing and pointing to an open newspaper. "The energetic Jones and the ubiquitous reporter have fixed it up between them. But you have had enough of the case. Better have your ham and eggs first."

---

1   Road in Victoria connecting Horseferry Road with Millbank (on the Thames).

I took the paper from him and read the short notice, which was headed "Mysterious Business at Upper Norwood."

"About twelve o'clock last night," said the *Standard*.[1] "Mr. Bartholomew Sholto, of Pondicherry Lodge, Upper Norwood, was found dead in his room under circumstances which point to foul play. As far as we can learn, no actual traces of violence were found upon Mr. Sholto's person, but a valuable collection of Indian gems which the deceased gentleman had inherited from his father has been carried off. The discovery was first made by Mr. Sherlock Holmes and Dr. Watson, who had called at the house with Mr. Thaddeus Sholto, brother of the deceased. By a singular piece of good fortune, Mr. Athelney Jones, the well-known member of the detective police force, happened to be at the Norwood Police Station, and was on the ground within half an hour of the first alarm. His trained and experienced faculties were at once directed towards the detection of the criminals, with the gratifying result that the brother, Thaddeus Sholto, has already been arrested, together with the housekeeper, Mrs. Bernstone, an Indian butler named Lal Rao, and a porter, or gatekeeper, named McMurdo. It is quite certain that the thief or thieves were well acquainted with the house, for Mr. Jones's well-known technical knowledge and his powers of minute observation have enabled him to prove conclusively that the miscreants could not have entered by the door or by the window but must have made their way across the roof of the building, and so through a trap-door into a room which communicated with that in which the body was found. This fact, which has been very clearly made out, proves conclusively that it was no mere haphazard burglary. The prompt and energetic action of the officers of the law shows the great advantage of the presence on such occasions of a single vigorous and masterful mind. We cannot but think that it supplies an argument to those who would wish to see our detectives more decentralized, and so brought into closer and more effective touch with the cases which it is their duty to investigate."

"Isn't it gorgeous!" said Holmes, grinning over his coffee cup. "What do you think of it?"

"I think that we have had a close shave ourselves of being arrested for the crime."

---

1  Launched in 1827, the *Standard* (now the *Evening Standard*) was a London tabloid newspaper that appeared in multiple daily editions and appealed to a populist readership.

"So do I. I wouldn't answer for our safety now if he should happen to have another of his attacks of energy."

At this moment there was a loud ring at the bell, and I could hear Mrs. Hudson, our landlady, raising her voice in a wail of expostulation and dismay.

"By heavens, Holmes," I said, half rising, "I believe that they are really after us."

"No, it's not quite so bad as that. It is the unofficial force—the Baker Street irregulars."

As he spoke, there came a swift pattering of naked feet upon the stairs, a clatter of high voices, and in rushed a dozen dirty and ragged little street Arabs. There was some show of discipline among them, despite their tumultuous entry, for they instantly drew up in line and stood facing us with expectant faces. One of their number, taller and older than the others, stood forward with an air of lounging superiority which was very funny in such a disreputable little scarecrow.

"Got your message, sir," said he, "and brought 'em on sharp. Three bob and a tanner[1] for tickets."

"Here you are," said Holmes, producing some silver. "In future they can report to you, Wiggins, and you to me. I cannot have the house invaded in this way. However, it is just as well that you should all hear the instructions. I want to find the whereabouts of a steam launch called the *Aurora*, owner Mordecai Smith, black with two red streaks, funnel black with a white band. She is down the river somewhere. I want one boy to be at Mordecai Smith's landing-stage opposite Millbank to say if the boat comes back. You must divide it out among yourselves and do both banks thoroughly. Let me know the moment you have news. Is that all clear?"

"Yes, guv'nor," said Wiggins.

"The old scale of pay, and a guinea to the boy who finds the boat. Here's a day in advance. Now off you go!"

He handed them a shilling each, and away they buzzed down the stairs, and I saw them a moment later streaming down the street.

"If the launch is above water they will find her," said Holmes as he rose from the table and lit his pipe. "They can go everywhere, see everything, overhear everyone. I expect to hear before evening that they have spotted her. In the meanwhile, we can do

---

1 I.e., 3s 6d.

nothing but await results. We cannot pick up the broken trail until we find either the *Aurora* or Mr. Mordecai Smith."

"Toby could eat these scraps, I dare say. Are you going to bed, Holmes?"

"No: I am not tired. I have a curious constitution. I never remember feeling tired by work, though idleness exhausts me completely. I am going to smoke and to think over this queer business to which my fair client has introduced us. If ever man had an easy task, this of ours ought to be. Wooden-legged men are not so common, but the other man must, I should think, be absolutely unique."

"That other man again!"

"I have no wish to make a mystery of him to you, anyway. But you must have formed your own opinion. Now, do consider the data. Diminutive footmarks, toes never fettered by boots, naked feet, stone-headed wooden mace, great agility, small poisoned darts. What do you make of all this?"

"A savage!" I exclaimed. "Perhaps one of those Indians who were the associates of Jonathan Small."

"Hardly that," said he. "When first I saw signs of strange weapons I was inclined to think so, but the remarkable character of the footmarks caused me to reconsider my views. Some of the inhabitants of the Indian Peninsula are small men, but none could have left such marks as that. The Hindoo proper has long and thin feet. The sandal-wearing Mohammedan has the great toe well separated from the others because the thong is commonly passed between. These little darts, too, could only be shot in one way. They are from a blow-pipe. Now, then, where are we to find our savage?"

"South America," I hazarded.

He stretched his hand up and took down a bulky volume from the shelf.

"This is the first volume of a gazetteer which is now being published. It may be looked upon as the very latest authority. What have we here? 'Andaman Islands, situated 340 miles to the north of Sumatra, in the Bay of Bengal.' Hum! hum! What's all this? Moist climate, coral reefs, sharks, Port Blair, convict barracks, Rutland Island, cottonwoods—Ah here we are! 'The aborigines of the Andaman Islands may perhaps claim the distinction of being the smallest race upon this earth, though some anthropologists prefer the Bushmen of Africa, the Digger Indians of

America, and the Terra del Fuegians.[1] The average height is rather below four feet, although many full-grown adults may be found who are very much smaller than this. They are a fierce, morose, and intractable people, though capable of forming most devoted friendships when their confidence has once been gained.' Mark that, Watson. Now, then listen to this. 'They are naturally hideous, having large, misshapen heads, small fierce eyes, and distorted features. Their feet and hands, however, are remarkably small. So intractable and fierce are they, that all the efforts of the British officials have failed to win them over in any degree. They have always been a terror to shipwrecked crews, braining the survivors with their stone-headed clubs or shooting them with their poisoned arrows.[2] These massacres are invariably concluded by a cannibal[3] feast.' Nice, amiable people, Watson! If this fellow had been left to his own unaided devices, this affair might have taken an even more ghastly turn. I fancy that, even as it is, Jonathan Small would give a good deal not to have employed him."

"But how came he to have so singular a companion?"

"Ah, that is more than I can tell. Since, however, we had already determined that Small had come from the Andamans, it is not so very wonderful that this islander should be with him. No doubt we shall know all about it in time. Look here, Watson; you look regularly done. Lie down there on the sofa, and see if I can put you to sleep."

---

1  The "gazetteer" that Holmes consults is wonderfully spurious and completely out of date. The indigenous population of Tierra del Fuego (who had almost been exterminated by the end of the nineteenth century) were not particularly short; the Bushmen (correctly termed Khoi-Sân) of Southern Africa, although slight, are far from being the shortest people on earth, and the "digger Indians" (correctly the Paiute) of Oregon, Nevada and Utah, were not noticeably shorter than other Native American groups. While the Andamanese are short, they are still not the shortest people on earth, and are certainly on average more than four feet tall. For the developments in scholarship about the Andamanese current at the time, see Appendix D.

2  In fact the Andamanese possessed neither stone clubs nor poisoned arrows; see Appendix D.

3  Despite many early travellers' tales associating the Andamanese with cannibalism, no evidence exists for them ever having practised it and all scholarship refutes this suspicion. See Appendix D.

He took up his violin from the corner, and as I stretched myself out he began to play some low, dreamy, melodious air—his own, no doubt, for he had a remarkable gift for improvisation.[1] I have a vague remembrance of his gaunt limbs, his earnest face, and the rise and fall of his bow. Then I seemed to be floated peacefully away upon a soft sea of sound, until I found myself in dreamland, with the sweet face of Mary Morstan looking down upon me.

## Chapter 9
## A Break in the Chain

It was late in the afternoon before I woke, strengthened and refreshed. Sherlock Holmes still sat exactly as I had left him, save that he had laid aside his violin and was deep in a book. He looked across at me as I stirred, and I noticed that his face was dark and troubled.

"You have slept soundly," he said. "I feared that our talk would wake you."

"I heard nothing," I answered. "Have you had fresh news, then?"

"Unfortunately, no. I confess that I am surprised and disappointed. I expected something definite by this time. Wiggins has just been up to report. He says that no trace can be found of the launch. It is a provoking check, for every hour is of importance."

"Can I do anything? I am perfectly fresh now, and quite ready for another night's outing."

"No; we can do nothing. We can only wait. If we go ourselves the message might come in our absence and delay be caused. You can do what you will, but I must remain on guard."

"Then I shall run over to Camberwell and call upon Mrs. Cecil Forrester. She asked me to, yesterday."

"On Mrs. Cecil Forrester?" asked Holmes, with the twinkle of a smile in his eyes.

"Well, of course on Miss Morstan, too. They were anxious to hear what happened."

---

1 One of the most famous fictional violinists, Holmes's improvisations have attracted much critical speculation. For a fascinating discussion of Holmes's apparent virtuosity, see the Danish violinist Jens Byskov Jensen's article, "The avant-garde Sherlock Holmes," in *The Baker Street Journal* (53:1): 13-18 at <http://www.bakerstreetjournal.com/images/Avant_Garde_Holmes_-_Jensen.pdf>.

"I would not tell them too much," said Holmes. "Women are never to be entirely trusted—not the best of them."

I did not pause to argue over this atrocious sentiment. *awareness of sexism*

"I shall be back in an hour or two," I remarked.

"All right! Good luck! But, I say, if you are crossing the river you may as well return Toby, for I don't think it is at all likely that we shall have any use for him now."

I took our mongrel accordingly and left him, together with a half-sovereign, at the old naturalist's in Pinchin Lane. At Camberwell I found Miss Morstan a little weary after her night's adventures but very eager to hear the news. Mrs. Forrester, too, was full of curiosity. I told them all that we had done, suppressing, however, the more dreadful parts of the tragedy. Thus, although I spoke of Mr. Sholto's death, I said nothing of the exact manner and method of it. With all my omissions, however, there was enough to startle and amaze them.

"It is a romance!" cried Mrs. Forrester. "An injured lady, half a million in treasure, a black cannibal, and a wooden-legged ruffian. They take the place of the conventional dragon or wicked earl."

"And two knight-errants to the rescue," added Miss Morstan with a bright glance at me.

"Why, Mary, your fortune depends upon the issue of this search. I don't think that you are nearly excited enough. Just imagine what it must be to be so rich and to have the world at your feet!"

It sent a little thrill of joy to my heart to notice that she showed no sign of elation at the prospect. On the contrary, she gave a toss of her proud head, as though the matter were one in which she took small interest.

"It is for Mr. Thaddeus Sholto that I am anxious," she said. "Nothing else is of any consequence; but I think that he has behaved most kindly and honourably throughout. It is our duty to clear him of this dreadful and unfounded charge."

It was evening before I left Camberwell, and quite dark by the time I reached home. My companion's book and pipe lay by his chair, but he had disappeared. I looked about in the hope of seeing a note, but there was none.

"I suppose that Mr. Sherlock Holmes has gone out," I said to Mrs. Hudson as she came up to lower the blinds.

"No, sir. He has gone to his room, sir. Do you know, sir," sinking her voice into an impressive whisper, "I am afraid for his health?"

"Why so, Mrs. Hudson?"

"Well, he's that strange, sir. After you was gone he walked and he walked, up and down, and up and down, until I was weary of the sound of his footstep. Then I heard him talking to himself and muttering, and every time the bell rang out he came on the stair-head, with 'What is that, Mrs. Hudson?' And now he has slammed off to his room, but I can hear him walking away the same as ever. I hope he's not going to be ill, sir. I ventured to say something to him about cooling medicine, but he turned on me, sir, with such a look that I don't know how ever I got out of the room."

"I don't think that you have any cause to be uneasy, Mrs. Hudson," I answered. "I have seen him like this before. He has some small matter upon his mind which makes him restless."

I tried to speak lightly to our worthy landlady, but I was myself somewhat uneasy when through the long night I still from time to time heard the dull sound of his tread, and knew how his keen spirit was chafing against this involuntary inaction.

At breakfast-time he looked worn and haggard, with a little fleck of feverish colour upon either cheek.

"You are knocking yourself up, old man," I remarked. "I heard you marching about in the night."

"No, I could not sleep," he answered. "This infernal problem is consuming me. It is too much to be balked by so petty an obstacle, when all else had been overcome. I know the men, the launch, everything; and yet I can get no news. I have set other agencies at work and used every means at my disposal. The whole river has been searched on either side, but there is no news, nor has Mrs. Smith heard of her husband. I shall come to the conclusion soon that they have scuttled the craft. But there are objections to that."

"Or that Mrs. Smith has put us on a wrong scent."

"No, I think that may be dismissed. I had inquiries made, and there is a launch of that description."

"Could it have gone up the river?"

"I have considered that possibility, too, and there is a search-party who will work up as far as Richmond.[1] If no news comes to-day, I shall start off myself tomorrow and go for the men rather than the boat. But surely, surely, we shall hear something."

We did not, however. Not a word came to us either from Wiggins or from the other agencies. There were articles in most

---

1  A leafy area straddling the Thames in southwest London, some 13 kilometres (8 miles) upstream from Westminster.

of the papers upon the Norwood tragedy. They all appeared to be rather hostile to the unfortunate Thaddeus Sholto. No fresh details were to be found, however, in any of them, save that an inquest was to be held upon the following day. I walked over to Camberwell in the evening to report our ill-success to the ladies, and on my return I found Holmes dejected and somewhat morose. He would hardly reply to my questions and busied himself all the evening in an abstruse chemical analysis which involved much heating of retorts and distilling of vapours, ending at last in a smell which fairly drove me out of the apartment. Up to the small hours of the morning I could hear the clinking of his test-tubes which told me that he was still engaged in his malodorous experiment.

In the early dawn I woke with a start, and was surprised to find him standing by my bedside, clad in a rude sailor dress with a pea-jacket, and a coarse red scarf round his neck.

"I am off down the river, Watson," said he. "I have been turning it over in my mind, and I can see only one way out of it. It is worth trying, at all events."

"Surely I can come with you, then?" said I.

"No; you can be much more useful if you will remain here as my representative. I am loath to go, for it is quite on the cards that some message may come during the day, though Wiggins was despondent about it last night. I want you to open all notes and telegrams, and to act on your own judgment if any news should come. Can I rely upon you?"

"Most certainly."

"I am afraid that you will not be able to wire to me, for I can hardly tell yet where I may find myself. If I am in luck, however, I may not be gone so very long. I shall have news of come sort or other before I get back."

I had heard nothing of him by breakfast time. On opening the *Standard*, however, I found that there was a fresh allusion to the business. "With reference to the Upper Norwood tragedy," it remarked, "we have reason to believe that the matter promises to be even more complex and mysterious than was originally supposed. Fresh evidence has shown that it is quite impossible that Mr. Thaddeus Sholto could have been in any way concerned in the matter. He and the housekeeper, Mrs. Bernstone, were both released yesterday evening. It is believed, however, that the police have a clue as to the real culprits, and that it is being prosecuted by Mr. Athelney Jones, of Scotland Yard, with all his well-known energy and sagacity. Further arrests may be expected at any moment."

"That is satisfactory so far as it goes," thought I. "Friend Sholto is safe, at any rate. I wonder what the fresh clue may be, though it seems to be a stereotyped form whenever the police have made a blunder."

I tossed the paper down upon the table, but at that moment my eye caught an advertisement in the agony column. It ran in this way:

"LOST—Whereas Mordecai Smith, boatman, and his son Jim, left Smith's Wharf at or about three o'clock last Tuesday morning in the steam launch *Aurora*, black with two red stripes, funnel black with a white band, the sum of five pounds will be paid to anyone who can give information to Mrs. Smith, at Smith's Wharf, or at 221*b*, Baker Street, as to the whereabouts of the said Mordecai Smith and the launch *Aurora*."

This was clearly Holmes's doing. The Baker Street address was enough to prove that. It struck me as rather ingenious, because it might be read by the fugitives without their seeing in it more than the natural anxiety of a wife for her missing husband.

It was a long day. Every time that a knock came to the door, or a sharp step passed in the street, I imagined that it was either Holmes returning or an answer to his advertisement. I tried to read, but my thoughts would wander off to our strange quest and to the ill-assorted and villainous pair whom we were pursuing. Could there be, I wondered, some radical flaw in my companion's reasoning? Might he not be suffering from some huge self-deception? Was it not possible that his nimble and speculative mind had built up this wild theory upon faulty premises? I had never known him to be wrong, and yet the keenest reasoner may occasionally be deceived. He was likely, I thought, to fall into error through the over-refinement of his logic—his preference for a subtle and bizarre explanation when a plainer and more common-place one lay ready to his hand. Yet, on the other hand, I had myself seen the evidence, and I had heard the reasons for his deductions. When I looked back on the long chain of curious circumstances, many of them trivial in themselves but all tending in the same direction, I could not disguise from myself that even if Holmes's explanation were incorrect the true theory must be equally *outré*[1] and startling.

---

1  Far-fetched (French).

At three o'clock on the afternoon there was a loud peal at the bell, an authoritative voice in the hall, and, to my surprise, no less a person than Mr. Athelney Jones was shown up to me. Very different was he, however, from the brusque and masterful professor of common sense who had taken over the case so confidently at Upper Norwood. His expression was downcast, and his bearing meek and even apologetic.

"Good-day, sir; good-day," said he. "Mr. Sherlock Holmes is out, I understand."

"Yes, and I cannot be sure when he will be back. But perhaps you would care to wait. Take that chair and try one of these cigars."

"Thank you; I don't mind if I do," said he, mopping his face with a red bandanna handkerchief.

"And a whisky and soda?"

"Well, half a glass. It is very hot for the time of year, and I have had a good deal to worry and try me. You know my theory about this Norwood case?"

"I remember that you expressed one."

"Well, I have been obliged to reconsider it. I had my net drawn tightly round Mr. Sholto, sir, when pop he went through a hole in the middle of it. He was able to prove an alibi which could not be shaken. From the time that he left his brother's room he was never out of sight of someone or other. So it could not be he who climbed over roofs and through trapdoors. It's a very dark case, and my professional credit is at stake. I should be very glad of a little assistance."

"We all need help sometimes," said I.

"Your friend, Mr. Sherlock Holmes, is a wonderful man, sir," said he in a husky and confidential voice, "He's a man who is not to be beat. I have known that young man go into a good many cases, but I never saw the case yet that he could not throw a light upon. He is irregular in his methods and a little quick perhaps in jumping at theories, but, on the whole, I think he would have made a most promising officer, and I don't care who knows it. I have had a wire from him this morning, by which I understand that he has got some clue to this Sholto business. Here is his message."

He took the telegram out of his pocket and handed it to me. It was dated from Poplar[1] at twelve o'clock. "Go to Baker Street

---

1  A working-class area of East London, on the north bank of the Thames opposite Greenwich.

at once," it said. "If I have not returned, wait for me. I am close on the track of the Sholto gang. You can come with us to-night if you want to be in at the finish."

"This sounds well. He has evidently picked up the scent again," said I.

"Ah, then he has been at fault too," exclaimed Jones with evident satisfaction. "Even the best of us are thrown off sometimes. Of course this may prove to be a false alarm; but it is my duty as an officer of the law to allow no chance to slip. But there is someone at the door. Perhaps this is he."

A heavy step was heard ascending the stair, with a great wheezing and rattling as from a man who was sorely put to it for breath. Once or twice he stopped, as though the climb were too much for him, but at last he made his way to our door and entered. His appearance corresponded to the sounds which we had heard. He was an aged man, clad in seafaring garb, with an old pea-jacket buttoned up to his throat. His back was bowed, his knees were shaky, and his breathing was painfully asthmatic. As he leaned upon a thick oaken cudgel his shoulders heaved in the effort to draw the air into his lungs. He had a coloured scarf round his chin, and I could see little of his face save a pair of keen dark eyes, overhung by bushy white brows and long gray side-whiskers. Altogether he gave me the impression of a respectable master mariner who had fallen into years and poverty.

"What is it, my man?" I asked.

He looked about him in the slow methodical fashion of old age.

"Is Mr. Sherlock Holmes here?" said he.

"No; but I am acting for him. You can tell me any message you have for him."

"It was to him himself I was to tell it," said he.

"But I tell you that I am acting for him. Was it about Morde-cai Smith's boat?"

"Yes. I knows well where it is. An' I knows where the men he is after are. An' I knows where the treasure is. I knows all about it."

"Then tell me, and I shall let him know."

"It was to him I was to tell it," he repeated with the petulant obstinacy of a very old man.

"Well, you must wait for him."

"No, no; I ain't goin' to lose a whole day to please no one. If Mr. Holmes ain't here, then Mr. Holmes must find it all out for himself. I don't care about the look of either of you, and I won't tell a word."

He shuffled towards the door, but Athelney Jones got in front of him.

"Wait a bit, my friend," said he. "You have important information, and you must not walk off. We shall keep you, whether you like or not, until our friend returns."

The old man made a little run towards the door, but, as Athelney Jones put his broad back up against it, he recognized the uselessness of resistance.

"Pretty sort o' treatment this!" he cried, stamping his stick. "I come here to see a gentleman, and you two, who I never saw in my life, seize me and treat me in this fashion!"

"You will be none the worse," I said. "We shall recompense you for the loss of your time. Sit over here on the sofa, and you will not have long to wait."

He came across sullenly enough and seated himself with his face resting on his hands. Jones and I resumed our cigars and our talk. Suddenly, however, Holmes's voice broke in upon us.

"I think that you might offer me a cigar too," he said.

We both started in our chairs. There was Holmes sitting close to us with an air of quiet amusement.

"Holmes!" I exclaimed. "You here! But where is the old man?"

"Here is the old man," said he, holding out a heap of white hair. "Here he is—wig, whiskers, eyebrows, and all. I thought my disguise was pretty good, but I hardly expected that it would stand that test."

"Ah, you rogue!" cried Jones, highly delighted. "You would have made an actor and a rare one. You had the proper workhouse cough, and those weak legs of yours are worth ten pound a week. I thought I knew the glint of your eye, though. You didn't get away from us so easily, you see."

"I have been working in that get-up all day," said he, lighting his cigar. "You see, a good many of the criminal classes begin to know me—especially since our friend here took to publishing some of my cases: so I can only go on the war-path under some simple disguise like this. You got my wire?"

"Yes; that was what brought me here."

"How has your case prospered?"

"It has all come to nothing. I have had to release two of my prisoners, and there is no evidence against the other two."

"Never mind. We shall give you two others in the place of them. But you must put yourself under my orders. You are welcome to all the official credit, but you must act on the lines that I point out. Is that agreed?"

"Entirely, if you will help me to the men."

"Well, then, in the first place I shall want, a fast police-boat—a steam launch—to be at the Westminster Stairs at seven o'clock."

"That is easily managed. There is always one about there, but I can step across the road and telephone to make sure."

"Then I shall want two staunch men, in case of resistance."

"There will be two or three in the boat. What else?"

"When we secure the men we shall get the treasure. I think that it would be a pleasure to my friend here to take the box round to the young lady to whom half of it rightfully belongs. Let her be the first to open it. Eh, Watson?"

"It would be a great pleasure to me."

"Rather an irregular proceeding," said Jones, shaking his head. "However, the whole thing is irregular, and I suppose we must wink at it. The treasure must afterwards be handed over to the authorities until after the official investigation."

"Certainly. That is easily managed. One other point. I should much like to have a few details about this matter from the lips of Jonathan Small himself. You know I like to work the details of my cases out. There is no objection to my having an unofficial interview with him, either here in my rooms or elsewhere, as long as he is efficiently guarded?"

"Well, you are master of the situation. I have had no proof yet of the existence of this Jonathan Small. However, if you can catch him, I don't see how I can refuse you an interview with him."

"That is understood, then?"

"Perfectly. Is there anything else?"

"Only that I insist upon your dining with us. It will be ready in half an hour. I have oysters and a brace of grouse,[1] with something a little choice in white wines.—Watson, you have never yet recognized my merits as a housekeeper."

---

1 Yet another indication that Doyle's original setting for the novel was in September, for oysters are traditionally only eaten between September and April, and the shooting season in England and Scotland always commences on 12 August. Grouse would always be shot from the wild, and never farmed. Holmes's choice of white wine is surely to accompany the oysters and *not* the grouse.

# Chapter 10
## The End of the Islander

Our meal was a merry one. Holmes could talk exceedingly well when he chose, and that night he did choose. He appeared to be in a state of nervous exaltation. I have never known him so brilliant. He spoke on a quick succession of subjects—on miracle plays, on medieval pottery, on Stradivarius violins, on the Buddhism of Ceylon, and on the warships of the future—handling each as though he had made a special study of it. His bright humour marked the reaction from his black depression of the preceding days. Athelney Jones proved to be a sociable soul in his hours of relaxation, and faced his dinner with the air of a *bon vivant*. For myself, I felt elated at the thought that we were nearing the end of our task, and I caught something of Holmes's gaiety. None of us alluded during dinner to the cause which had brought us together.

When the cloth was cleared Holmes glanced at his watch, and filled up three glasses with port.

"One bumper," said he, "to the success of our little expedition. And now it is high time we were off. Have you a pistol, Watson?"

"I have my old service-revolver in my desk."

"You had best take it, then. It is well to be prepared. I see that the cab is at the door. I ordered it for half-past six."

It was a little past seven before we reached the Westminster wharf, and found our launch awaiting us. Holmes eyed it critically.

"Is there anything to mark it as a police-boat?"

"Yes, that green lamp at the side."

"Then take it off."

The small change was made, we stepped on board, and the ropes were cast off. Jones, Holmes, and I sat in the stern. There was one man at the rudder, one to tend the engines, and two burly police-inspectors forward.

"Where to?" asked Jones.

"To the Tower.[1] Tell them to stop opposite to Jacobson's Yard."

Our craft was evidently a very fast one. We shot past the long lines of loaded barges as though they were stationary. Holmes smiled with satisfaction as we overhauled a river steamer and left her behind us.

---

1 Located on the north bank of the Thames and on the easternmost edge of the City of London, the Tower of London was originally built in 1078 and, since 1303, has housed the Crown Jewels of England.

"We ought to be able to catch anything on the river," he said.

"Well, hardly that. But there are not many launches to beat us."

"We shall have to catch the *Aurora*, and she has a name for being a clipper. I will tell you how the land lies, Watson. You recollect how annoyed I was at being baulked by so small a thing?"

"Yes."

"Well, I gave my mind a thorough rest by plunging into a chemical analysis. One of our greatest statesmen has said that a change of work is the best rest. So it is. When I had succeeded in dissolving the hydrocarbon which I was at work at, I came back to our problem of the Sholtos, and thought the whole matter out again. My boys had been up the river and down the river without result. The launch was not at any landing-stage or wharf, nor had it returned. Yet it could hardly have been scuttled to hide their traces, though that always remained as a possible hypothesis if all else failed. I knew that this man Small had a certain degree of low cunning, but I did not think him capable of anything in the nature of delicate finesse. That is usually a product of higher education. I then reflected that since he had certainly been in London some time—as we had evidence that he maintained a continual watch over Pondicherry Lodge—he could hardly leave at a moment's notice, but would need some little time, if it were only a day, to arrange his affairs. That was the balance of probability, at any rate."

"It seems to me to be a little weak," said I; "it is more probable that he had arranged his affairs before ever he set out upon his expedition."

"No, I hardly think so. This lair of his would be too valuable a retreat in case of need for him to give it up until he was sure that he could do without it. But a second consideration struck me. Jonathan Small must have felt that the peculiar appearance of his companion, however much he may have top coated him, would give rise to gossip, and possibly be associated with this Norwood tragedy. He was quite sharp enough to see that. They had started from their headquarters under cover of darkness, and he would wish to get back before it was broad light. Now, it was past three o'clock, according to Mrs. Smith, when they got the boat. It would be quite bright, and people would be about in an hour or so. Therefore, I argued, they did not go very far. They paid Smith well to hold his tongue, reserved his launch for the final escape, and hurried to their lodgings with the treasure-box. In a couple of nights, when they had time to see what view the papers took, and whether there was any suspicion, they would make their way under cover of darkness to some ship at Gravesend or in the Downs, where no doubt they had already arranged for passages to America or the Colonies."

"But the launch? They could not have taken that to their lodgings."

"Quite so. I argued that the launch must be no great way off, in spite of its invisibility. I then put myself in the place of Small and looked at it as a man of his capacity would. He would probably consider that to send back the launch or to keep it at a wharf would make pursuit easy if the police did happen to get on his track. How, then, could he conceal the launch and yet have her at hand when wanted? I wondered what I should do myself if I were in his shoes. I could only think of one way of doing it. I might hand the launch over to some boat-builder or repairer, with directions to make a trifling change in her. She would then be removed to his shed or yard, and so be effectually concealed, while at the same time I could have her at a few hours' notice."

"That seems simple enough."

"It is just these very simple things which are extremely liable to be overlooked. However, I determined to act on the idea. I started at once in this harmless seaman's rig and inquired at all the yards down the river. I drew blank at fifteen, but at the sixteenth—Jacobson's—I learned that the *Aurora* had been handed over to them two days ago by a wooden-legged man, with some trivial directions as to her rudder. 'There ain't naught amiss with her rudder,' said the foreman. 'There she lies, with the red streaks.' At that moment who should come down but Mordecai Smith, the missing owner? He was rather the worse for liquor. I should not, of course, have known him, but he bellowed out his name and the name of his launch. 'I want her to-night at eight o'clock,' said he—'eight o'clock sharp, mind, for I have two gentlemen who won't be kept waiting.' They had evidently paid him well, for he was very flush of money, chucking shillings about to the men. I followed him some distance, but he subsided into an alehouse; so I went back to the yard, and, happening to pick up one of my boys on the way, I stationed him as a sentry over the launch. He is to stand at the water's edge and wave his handkerchief to us when they start. We shall be lying off in the stream, and it will be a strange thing if we do not take men, treasure, and all."

"You have planned it all very neatly, whether they are the right men or not," said Jones; "but if the affair were in my hands I should have had a body of police in Jacobson's Yard and arrested them when they came down."

"Which would have been never. This man Small is a pretty shrewd fellow. He would send a scout on ahead, and if anything made him suspicious he would lie snug for another week."

"But you might have stuck to Mordecai Smith, and so been led to their hiding-place," said I.

"In that case I should have wasted my day. I think that it is a hundred to one against Smith knowing where they live. As long as he has liquor and good pay, why should he ask questions? They send him messages what to do. No, I thought over every possible course, and this is the best."

While this conversation had been proceeding, we had been shooting the long series of bridges which span the Thames. As we passed the City the last rays of the sun were gilding the cross upon the summit of St. Paul's.[1] It was twilight before we reached the Tower.

"That is Jacobson's Yard," said Holmes, pointing to a bristle of masts and rigging on the Surrey side. "Cruise gently up and down here under cover of this string of lighters." He took a pair of night-glasses from his pocket and gazed some time at the shore. "I see my sentry at his post," he remarked, "but no sign of a handkerchief."

"Suppose we go downstream a short way and lie in wait for them," said Jones eagerly.

We were all eager by this time, even the policemen and stokers, who had a very vague idea of what was going forward.

"We have no right to take anything for granted," Holmes answered. "It is certainly ten to one that they go downstream, but we cannot be certain. From this point we can see the entrance of the yard, and they can hardly see us. It will be a clear night and plenty of light. We must stay where we are. See how the folk swarm over yonder in the gaslight."

"They are coming from work in the yard."

"Dirty-looking rascals, but I suppose every one has some little immortal spark concealed about him. You would not think it, to look at them. There is no *a priori* probability about it. A strange enigma is man!"

"Someone calls him a soul concealed in an animal," I suggested.

---

1 The present St. Paul's Cathedral, the fifth, was designed by Christopher Wren (1632-1723) and built between 1675 and 1710. It has been a site of Christian worship since the seventh century.

"Winwood Reade is good upon the subject," said Holmes. "He remarks that, while the individual man is an insoluble puzzle, in the aggregate he becomes a mathematical certainty. You can, for example, never foretell what any one man will do, but you can say with precision what an average number will be up to. Individuals vary, but percentages remain constant. So says the statistician. But do I see a handkerchief? Surely there is a white flutter over yonder."

"Yes, it is your boy," I cried. "I can see him plainly."

"And there is the *Aurora*," exclaimed Holmes, "and going like the devil! Full speed ahead, engineer. Make after that launch with the yellow light. By heaven, I shall never forgive myself if she proves to have the heels of us!"

She had slipped unseen through the yard-entrance and passed between two or three small craft, so that she had fairly got her speed up before we saw her. Now she was flying down the stream, near in to the shore, going at a tremendous rate. Jones looked gravely at her and shook his head.

"She is very fast," he said. "I doubt if we shall catch her."

"We *must* catch her!" cried Holmes, between his teeth. "Heap it on, stokers! Make her do all she can! If we burn the boat we must have them!"

We were fairly after her now. The furnaces roared, and the powerful engines whizzed and clanked, like a great metallic heart. Her sharp, steep prow cut through the still river-water and sent two rolling waves to right and to left of us. With every throb of the engines we sprang and quivered like a living thing. One great yellow lantern in our bows threw a long, flickering funnel of light in front of us. Right ahead a dark blur upon the water showed where the *Aurora* lay, and the swirl of white foam behind her spoke of the pace at which she was going We flashed past barges, steamers, merchant-vessels, in and out, behind this one and round the other. Voices hailed us out of the darkness, but still the *Aurora* thundered on, and still we followed close upon her track.

"Pile it on, men, pile it on!" cried Holmes, looking down into the engine-room, while the fierce glow from below beat upon his eager, aquiline face. "Get every pound of steam you can."

"I think we gain a little," said Jones with his eyes on the *Aurora*.

"I am sure of it," said I. "We shall be up with her in a very few minutes."

At that moment, however, as our evil fate would have it, a tug with three barges in tow blundered in between us. It was only by putting our helm hard down that we avoided a collision, and before we could round them and recover our way the *Aurora* had gained a good two hundred yards. She was still, however, well in view, and the murky, uncertain twilight was settling into a clear, starlit night. Our boilers were strained to their utmost, and the frail shell vibrated and creaked with the fierce energy which was driving us along. We had shot through the pool, past the West India Docks,[1] down the long Deptford Reach,[2] and up again after rounding the Isle of Dogs.[3] The dull blur in front of us resolved itself now clearly into the dainty *Aurora*. Jones turned our searchlight upon her, so that we could plainly see the figures upon her deck. One man sat by the stern, with something black between his knees, over which he stooped. Beside him lay a dark mass, which looked like a Newfoundland dog. The boy held the tiller, while against the red glare of the furnace I could see old Smith, stripped to the waist, and shovelling coals for dear life. They may have had some doubt at first as to whether we were really pursuing them, but now as we followed every winding and turning which they took there could no longer be any question about it. At Greenwich we were about three hundred paces behind them. At Blackwall[4] we could not have been more than two hundred and fifty. I have coursed many creatures in many countries during my checkered career, but never did sport give me such a wild thrill as this mad, flying man-hunt down the Thames. Steadily we drew in upon them, yard by yard. In the silence of the night we could hear the panting and clanking of their machinery. The man in the stern still crouched upon the

1   A series of three docks designed by Robert Milligan (1746-1809) and constructed between 1800 and 1802 on the Isle of Dogs in East London. As the name suggests, the docks were built specifically to service the trade in plantation goods (especially sugar) from the West Indies. After falling into dereliction in the twentieth century, the area has now been successfully redeveloped as a financial district, and rebranded under the name of London Docklands.
2   Deptford is on the south bank of the Thames, opposite Millwall on the Isle of Dogs. Tsar Peter the Great (1672-1725) of Russia lived in Deptford in 1698 as a student of shipbuilding.
3   A peninsula in East London surrounded on three sides by the Thames; its etymology is still disputed. The trajectory of the *Aurora* is downstream.
4   Downstream from Deptford and north of Greenwich, at the time Blackwall was the centre for shipbuilding on the Thames. The name is now synonymous with the two tunnels under the Thames, the first of which, constructed between 1892 and 1897, was at the time the longest underwater tunnel in the world.

deck, and his arms were moving as though he were busy, while every now and then he would look up and measure with a glance the distance which still separated us. Nearer we came and nearer. Jones yelled to them to stop. We were not more than four boat's-lengths behind them, both boats flying at a tremendous pace. It was a clear reach of the river, with Barking Level[1] upon one side and the melancholy Plumstead Marshes[2] upon the other. At our hail the man in the stern sprang up from the deck and shook his two clenched fists at us, cursing the while in a high, cracked voice. He was a good-sized, powerful man, and as he stood poising himself with legs astride, I could see that from the thigh downward there was but a wooden stump upon the right side. At the sound of his strident, angry cries, there was movement in the huddled bundle upon the deck. It straightened itself into a little black man—the smallest I have ever seen—with a great, mis-shapen head and a shock of tangled, dishevelled hair. Holmes had already drawn his revolver, and I whipped out mine at the sight of this savage, distorted creature. He was wrapped in some sort of dark ulster or blanket, which left only his face exposed, but that face was enough to give a man a sleepless night. Never have I seen features so deeply marked with all bestiality and cruelty. His small eyes glowed and burned with a sombre light, and his thick lips were writhed back from his teeth, which grinned and chattered at us with half animal fury.

"Fire if he raises his hand," said Holmes quietly.

We were within a boat's-length by this time, and almost within touch of our quarry. I can see the two of them now as they stood, the white man with his legs far apart, shrieking out curses, and the unhallowed dwarf with his hideous face, and his strong yellow teeth gnashing at us in the light of our lantern

It was well that we had so clear a view of him. Even as we looked he plucked out from under his covering a short, round piece of wood, like a school-ruler, and clapped it to his lips. Our pistols rang out together. He whirled round, threw up his arms and, with a kind of choking cough, fell sideways into the stream. I caught one glimpse of his venomous, menacing eyes amid the white swirl of the waters. At the same moment the wooden-

---

1 Further downstream from Blackwall on the north shore of the Thames, and originally the centre of London's fishing industry; by the 1880s pollution had destroyed the fisheries.

2 On the south side of the Thames, opposite Barking. The eastern part of Plumstead was occupied by the Royal Arsenal.

legged man threw himself upon the rudder and put it hard down so that his boat made straight in for the southern bank, while we shot past her stern, only clearing her by a few feet. We were round after her in an instant, but she was already nearly at the bank. It was a wild and desolate place, where the moon glimmered upon a wide expanse of marsh-land, with pools of stagnant water and beds of decaying vegetation. The launch, with a dull thud, ran up upon the mud-bank, with her bow in the air and her stern flush with the water. The fugitive sprang out, but his stump instantly sank its whole length into the sodden soil. In vain he struggled and writhed. Not one step could he possibly take either forward or backward. He yelled in impotent rage, and kicked frantically into the mud with his other foot; but his struggles only bored his wooden pin the deeper into the sticky bank. When we brought our launch alongside he was so firmly anchored that it was only by throwing the end of a rope over his shoulders that we were able to haul him out, and to drag him, like some evil fish, over our side. The two Smiths, father and son, sat sullenly in their launch but came aboard meekly enough when commanded. The *Aurora* herself we hauled off and made fast to our stern. A solid iron chest of Indian workmanship stood upon the deck. This, there could be no question, was the same that had contained the ill-omened treasure of the Sholtos. There was no key, but it was of considerable weight, so we transferred it carefully to our own little cabin. As we steamed slowly upstream again, we flashed our searchlight in every direction, but there was no sign of the Islander. Somewhere in the dark ooze at the bottom of the Thames lie the bones of that strange visitor to our shores.

"See here," said Holmes, pointing to the wooden hatchway. "We were hardly quick enough with our pistols." There, sure enough, just behind where we had been standing, stuck one of those murderous darts which we knew so well. It must have whizzed between us at the instant we fired. Holmes smiled at it and shrugged his shoulders in his easy fashion, but I confess that it turned me sick to think of the horrible death which had passed so close to us that night.

Chapter 11
The Great Agra Treasure

Our captive sat in the cabin opposite to the iron box which he had done so much and waited so long to gain. He was a sun-burned, reckless-eyed fellow, with a network of lines and wrinkles all over his mahogany features, which told of a hard, open-air life. There was a singular prominence about his bearded chin which marked a man who was not to be easily turned from his purpose. His age may have been fifty or thereabouts, for his black, curly hair was thickly shot with gray. His face in repose was not an unpleasing one, though his heavy brows and aggressive chin gave him, as I had lately seen, a terrible expression when moved to anger. He sat now with his handcuffed hands upon his lap, and his head sunk upon his breast, while he looked with his keen, twinkling eyes at the box which had been the cause of his ill-doings. It seemed to me that there was more sorrow than anger in his rigid and contained countenance. Once he looked up at me with a gleam of something like humour in his eyes.

"Well, Jonathan Small," said Holmes, lighting a cigar, "I am sorry that it has come to this."

"And so am I, sir," he answered frankly. "I don't believe that I can swing over the job. I give you my word on the book that I never raised hand against Mr. Sholto. It was that little hell-hound Tonga who shot one of his cursed darts into him. I had no part in it, sir. I was as grieved as if it had been my blood-relation. I welted the little devil with the slack end of the rope for it, but it was done, and I could not undo it again."

"Have a cigar," said Holmes; "and you had best take a pull out of my flask, for you are very wet. How could you expect so small and weak a man as this black fellow to overpower Mr. Sholto and hold him while you were climbing the rope?"

"You seem to know as much about it as if you were there, sir. The truth is that I hoped to find the room clear. I knew the habits of the house pretty well, and it was the time when Mr. Sholto usually went down to his supper.[1] I shall make no secret of the business. The best defence that I can make is just the simple truth. Now, if it had been the old major I would have swung for him with a light heart. I would have thought no more of knifing him than of smoking this cigar. But it's cursed hard that I should be lagged[2] over this young Sholto, with whom I had no quarrel whatever."

"You are under the charge of Mr. Athelney Jones, of Scotland Yard. He is going to bring you up to my rooms, and I shall ask

---

1 Another inconsistency in Small's account of the murder, for before the murder Bartholomew Sholto was visited by his brother Thaddeus, who did not leave until 10 pm, and heard the door locked from the inside on his way out. Small and Tonga arrived after Thaddeus had left.

2 Caught.

you for a true account of the matter. You must make a clean breast of it, for if you do I hope that I may be of use to you. I think I can prove that the poison acts so quickly that the man was dead before ever you reached the room."

"That he was, sir. I never got such a turn in my life as when I saw him grinning at me with his head on his shoulder as I climbed through the window. It fairly shook me, sir. I'd have half killed Tonga for it if he had not scrambled off. That was how he came to leave his club, and some of his darts too, as he tells me, which I dare say helped to put you on our track; though how you kept on it is more than I can tell. I don't feel no malice against you for it. But it does seem a queer thing," he added with a bitter smile, "that I, who have a fair claim to half a million of money, should spend the first half of my life building a breakwater in the Andamans, and am like to spend the other half digging drains at Dartmoor.[1] It was an evil day for me when first I clapped eyes upon the merchant Achmet and had to do with the Agra treasure, which never brought anything but a curse yet upon the man who owned it. To him it brought murder, to Major Sholto it brought fear and guilt, to me it has meant slavery for life."

At this moment Athelney Jones thrust his broad face and heavy shoulders into the tiny cabin.

"Quite a family party," he remarked. "I think I shall have a pull at that flask, Holmes. Well, I think we may all congratulate each other. Pity we didn't take the other alive, but there was no choice. I say, Holmes, you must confess that you cut it rather fine. It was all we could do to overhaul her."

"All is well that ends well," said Holmes. "But I certainly did not know that the *Aurora* was such a clipper."

"Smith says she is one of the fastest launches on the river, and that if he had had another man to help him with the engines we should never have caught her. He swears he knew nothing of this Norwood business."

"Neither he did," cried our prisoner—"not a word. I chose his launch because I heard that she was a flier. We told him nothing;

---

1   A remote and dramatically beautiful area of wilderness in the centre of Devon, Dartmoor was synonymous with its prison, built between 1806 and 1809 and still in use today. Originally built to house prisoners of war from the Napoleonic Wars, Dartmoor Prison became a civilian institution in 1851. Doyle's *The Hound of the Baskervilles* (1901-02) is largely set in Dartmoor.

but we paid him well, and he was to get something handsome if we reached our vessel, the *Esmeralda*,[1] at Gravesend, outward bound for the Brazils."

"Well, if he has done no wrong we shall see that no wrong comes to him. If we are pretty quick in catching our men, we are not so quick in condemning them." It was amusing to notice how the consequential Jones was already beginning to give himself airs on the strength of the capture. From the slight smile which played over Sherlock Holmes's face, I could see that the speech had not been lost upon him.

"We will be at Vauxhall Bridge presently," said Jones, "and shall land you, Dr. Watson, with the treasure-box. I need hardly tell you that I am taking a very grave responsibility upon myself in doing this. It is most irregular; but of course an agreement is an agreement. I must, however, as a matter of duty, send an inspector with you, since you have so valuable a charge. You will drive, no doubt?"

"Yes, I shall drive."

"It is a pity there is no key, that we may make an inventory first. You will have to break it open. Where is the key, my man?"

"At the bottom of the river," said Small shortly.

"Hum! There was no use your giving this unnecessary trouble. We have had work enough already through you. However, doctor, I need not warn you to be careful. Bring the box back with you to the Baker Street rooms. You will find us there, on our way to the station."

They landed me at Vauxhall, with my heavy iron box, and with a bluff, genial inspector as my companion. A quarter of an hour's drive brought us to Mrs. Cecil Forrester's. The servant seemed surprised at so late a visitor. Mrs. Cecil Forrester was out for the evening, she explained, and likely to be very late. Miss Morstan, however, was in the drawing-room, so to the drawing-room I went, box in hand, leaving the obliging inspector in the cab.

She was seated by the open window, dressed in some sort of white diaphanous material, with a little touch of scarlet at the neck and waist. The soft light of a shaded lamp fell upon her as she leaned back in the basket chair, playing over her sweet grave face, and tinting with a dull, metallic sparkle the rich coils of her luxuriant hair. One white arm and hand drooped over the side of the chair, and her whole pose and figure spoke of an absorbing melancholy. At the sound of my footfall she sprang to her feet,

---

1  The Portuguese and Spanish word for emerald.

however, and a bright flush of surprise and of pleasure coloured her pale cheeks.

"I heard a cab drive up," she said. "I thought that Mrs. Forrester had come back very early, but I never dreamed that it might be you. What news have you brought me?"

"I have brought something better than news," said I, putting down the box upon the table and speaking jovially and boisterously, though my heart was heavy within me. "I have brought you something which is worth all the news in the world. I have brought you a fortune."

She glanced at the iron box.

"Is that the treasure then?" she asked, coolly enough.

"Yes, this is the great Agra treasure. Half of it is yours and half is Thaddeus Sholto's. You will have a couple of hundred thousand each. Think of that! An annuity of ten thousand pounds. There will be few richer young ladies in England. Is it not glorious?"

I think I must have been rather over-acting my delight, and that she detected a hollow ring in my congratulations, for I saw her eyebrows rise a little, and she glanced at me curiously.

"If I have it," said she, "I owe it to you."

"No, no," I answered, "not to me but to my friend Sherlock Holmes. With all the will in the world, I could never have followed up a clue which has taxed even his analytical genius. As it was, we very nearly lost it at the last moment."

"Pray sit down and tell me all about it, Dr. Watson," said she.

I narrated briefly what had occurred since I had seen her last. Holmes's new method of search, the discovery of the *Aurora*, the appearance of Athelney Jones, our expedition in the evening, and the wild chase down the Thames. She listened with parted lips and shining eyes to my recital of our adventures. When I spoke of the dart which had so narrowly missed us, she turned so white that I feared that she was about to faint.

"It is nothing," she said as I hastened to pour her out some water. "I am all right again. It was a shock to me to hear that I had placed my friends in such horrible peril."

"That is all over," I answered. "It was nothing. I will tell you no more gloomy details. Let us turn to something brighter. There is the treasure. What could be brighter than that? I got leave to bring it with me, thinking that it would interest you to be the first to see it."

"It would be of the greatest interest to me," she said. There was no eagerness in her voice, however. It had struck her, doubtless, that it might seem ungracious upon her part to be indifferent to a prize which had cost so much to win.

"What a pretty box!" she said, stooping over it. "This is Indian work, I suppose?"

"Yes; it is Benares[1] metal-work."

"And so heavy!" she exclaimed, trying to raise it. "The box alone must be of some value. Where is the key?"

"Small threw it into the Thames," I answered. "I must borrow Mrs. Forrester's poker."

There was in the front a thick and broad hasp, wrought in the image of a sitting Buddha.[2] Under this I thrust the end of the poker and twisted it outward as a lever. The hasp sprang open with a loud snap. With trembling fingers I flung back the lid. We both stood gazing in astonishment. The box was empty!

No wonder that it was heavy. The ironwork was two-thirds of an inch thick all round. It was massive, well made, and solid, like a chest constructed to carry things of great price, but not one shred or crumb of metal or jewellery lay within it. It was absolutely and completely empty.

"The treasure is lost," said Miss Morstan calmly.

As I listened to the words and realized what they meant, a great shadow seemed to pass from my soul. I did not know how this Agra treasure had weighed me down, until now that it was finally removed. It was selfish, no doubt, disloyal, wrong, but I could realize nothing save that the golden barrier was gone from between us.

"Thank God!" I ejaculated from my very heart.

She looked at me with a quick, questioning smile.

"Why do you say that?" she asked.

"Because you are within my reach again," I said, taking her hand. She did not withdraw it. "Because I love you, Mary, as

---

1 Benares, now called Varanasi and also known as Kashi, is the holiest city of Hinduism and is located on the banks of the Ganges some 570 kilometres (350 miles) east of Agra. Benares is best known for producing some of the finest silk saris in the world, though it does produce some excellent brassware as well.

2 Despite his own relative ignorance about the subject, Doyle was prepared to pander to the growing popular interest in Buddhism at the time, fuelled by books such as A.P. Sinnett's *Esoteric Buddhism* (1883) and Arthur Lillie's *Koot Hoomi Unveiled; or Tibetan "Buddhists" versus the Buddhists of Tibet* (1884); his earlier *The Mystery of Cloomber* (1889) had featured Buddhist monks. The idea of a Buddha image forming the fastening for a treasure box is both highly irreverent and irregular.

truly as ever a man loved a woman. Because this treasure, these riches, sealed my lips. Now that they are gone I can tell you how I love you. That is why I said, 'Thank God.'"

"Then I say 'Thank God,' too," she whispered as I drew her to my side.

Whoever had lost a treasure, I knew that night that I had gained one.

## Chapter 12
### The Strange Story of Jonathan Small

A very patient man was that inspector in the cab, for it was a weary time before I rejoined him. His face clouded over when I showed him the empty box.

"There goes the reward!" said he gloomily. "Where there is no money there is no pay. This night's work would have been worth a tenner each to Sam Brown and me if the treasure had been there."

"Mr. Thaddeus Sholto is a rich man," I said; "he will see that you are rewarded, treasure or no."

The inspector shook his head despondently, however.

"It's a bad job," he repeated; "and so Mr. Athelney Jones will think."

His forecast proved to be correct, for the detective looked blank enough when I got to Baker Street and showed him the empty box. They had only just arrived, Holmes, the prisoner, and he, for they had changed their plans so far as to report themselves at a station upon the way. My companion lounged in his armchair with his usual listless expression, while Small sat stolidly opposite to him with his wooden leg cocked over his sound one. As I exhibited the empty box he leaned back in his chair and laughed aloud.

"This is your doing, Small," said Athelney Jones angrily.

"Yes, I have put it away where you shall never lay hand upon it," he cried exultantly. "It is my treasure, and if I can't have the loot[1] I'll take darned good care that no one else does. I tell you that no living man has any right to it, unless it is three men who are in the Andaman convict-barracks and myself. I know now that I cannot have the use of it, and I know that they cannot. I have acted all through for them as much as for myself. It's been the sign of four with us always. Well, I know that they would have

---

1  Suitably enough, an Anglo-Indian word with a Hindi stem, lūt.

had me do just what I have done, and throw the treasure into the Thames rather than let it go to kith or kin of Sholto or Morstan. It was not to make them rich that we did for Achmet. You'll find the treasure where the key is and where little Tonga is. When I saw that your launch must catch us, I put the loot away in a safe place. There are no rupees for you this journey."

"You are deceiving us, Small," said Athelney Jones sternly; "if you had wished to throw the treasure into the Thames, it would have been easier for you to have thrown box and all."

"Easier for me to throw and easier for you to recover," he answered with a shrewd, side-long look. "The man that was clever enough to hunt me down is clever enough to pick an iron box from the bottom of a river. Now that they are scattered over five miles or so, it may be a harder job. It went to my heart to do it, though. I was half mad when you came up with us. However, there's no good grieving over it. I've had ups in my life, and I've had downs, but I've learned not to cry over spilled milk."

"This is a very serious matter, Small," said the detective. "If you had helped justice, instead of thwarting it in this way, you would have had a better chance at your trial."

"Justice!" snarled the ex-convict. "A pretty justice! Whose loot is this, if it is not ours? Where is the justice that I should give it up to those who have never earned it? Look how I have earned it! Twenty long years in that fever-ridden swamp, all day at work under the mangrove-tree, all night chained up in the filthy convict-huts, bitten by mosquitoes, racked with ague, bullied by every cursed black-faced policeman[1] who loved to take it out of a white man. That was how I earned the Agra treasure, and you talk to me of justice because I cannot bear to feel that I have paid this price only that another may enjoy it! I would rather swing a score of times, or have one of Tonga's darts in my hide, than live in a convict's cell and feel that another man is at his ease in a palace with the money that should be mine."

Small had dropped his mask of stoicism, and all this came out in a wild whirl of words, while his eyes blazed, and the handcuffs clanked together with the impassioned movement of his hands. I

1  The Andaman penal settlement used both self-policing, where trustees would police their fellow inmates, and traditional civilian police. After the murder of Lord Mayo brought a new wave of criticism of the perceived laxity of the system, the authorities started recruiting a new Indian civil police, largely from the Punjab and the North-West Provinces. See Appendices D5 and D6.

could understand, as I saw the fury and the passion of the man, that it was no groundless or unnatural terror which had possessed Major Sholto when he first learned that the injured convict was upon his track.

"You forget that we know nothing of all this," said Holmes quietly. "We have not heard your story, and we cannot tell how far justice may originally have been on your side."

"Well, sir, you have been very fair-spoken to me, though I can see that I have you to thank that I have these bracelets upon my wrists. Still, I bear no grudge for that. It is all fair and above-board. If you want to hear my story, I have no wish to hold it back. What I say to you is God's truth, every word of it. Thank you, you can put the glass beside me here, and I'll put my lips to it if I am dry.

"I am a Worcestershire man myself, born near Pershore.[1] I dare say you would find a heap of Smalls living there now if you were to look. I have often thought of taking a look round there, but the truth is that I was never much of a credit to the family, and I doubt if they would be so very glad to see me. They were all steady, chapel-going folk, small farmers, well known and respected over the country-side, while I was always a bit of a rover. At last, however, when I was about eighteen, I gave them no more trouble, for I got into a mess over a girl and could only get out of it again by taking the Queen's shilling and joining the 3rd Buffs, which was just starting for India.

"I wasn't destined to do much soldiering, however. I had just got past the goose-step and learned to handle my musket, when I was fool enough to go swimming in the Ganges. Luckily for me, my company sergeant, John Holder, was in the water at the same time, and he was one of the finest swimmers in the service. A crocodile took me, just as I was half-way across, and nipped off my right leg as clean as a surgeon could have done it, just above the knee.[2] What with the shock and the loss of blood, I fainted, and should have been drowned if Holder had not caught hold of me and paddled for the bank. I was five months in hospital over it, and when at last I was able to limp out of it with this timber toe strapped to my stump, I found myself invalided out of the Army and unfitted for any active occupation.

---

1  A small market town on the banks of the River Avon in Worcestershire.
2  A highly implausible account, for crocodiles favour drowning their victims.

"I was, as you can imagine, pretty down on my luck at this time, for I was a useless cripple, though not yet in my twentieth year. However, my misfortune soon proved to be a blessing in disguise. A man named Abel White, who had come out there as an indigo-planter,[1] wanted an overseer to look after his coolies and keep them up to their work. He happened to be a friend of our colonel's, who had taken an interest in me since the accident. To make a long story short, the colonel recommended me strongly for the post, and, as the work was mostly to be done on horseback, my leg was no great obstacle, for I had enough thigh left to keep a good grip on the saddle. What I had to do was to ride over the plantation, to keep an eye on the men as they worked, and to report the idlers. The pay was fair, I had comfortable quarters, and altogether I was content to spend the remainder of my life in indigo-planting. Mr. Abel White was a kind man, and he would often drop into my little shanty and smoke a pipe with me, for white folk out there feel their hearts warm to each other as they never do here at home.

"Well, I was never in luck's way long. Suddenly, without a note of warning, the great mutiny broke upon us. One month India lay as still and peaceful, to all appearance, as Surrey or Kent; the next there were two hundred thousand black devils let loose, and the country was a perfect hell. Of course you know all about it, gentlemen—a deal more than I do, very like, since reading is not in my line. I only know what I saw with my own eyes. Our plan-

---

1   One of the main cash crops in the colonial period, indigo was a taxed item and its production was tightly controlled by the British. Until the invention of artificial industrial dyeing, indigo was the most widely used dye for fabrics. Throughout the nineteenth century, Indian peasants were frequently forced to cultivate indigo rather than rice or wheat, in order to produce taxable revenue for the British. Resentment over the forced cultivation of indigo was another of the grievances contributing to the "Mutiny" of 1857-58. Mathura did indeed have an indigo plantation, the Oomerghur Concern, which was owned by the firm of Mackillop, Stewart and Co., and managed by a resident British manager, Thomas Churcher; the plantation survived the "Mutiny," albeit with a new resident manager. Indigo dominated commercial activity in the East India Company's territory in this period; 26 of the 36 pages listing "manufactories" in *The New Calcutta Directory of the town of Calcutta, Bengal, the North-West Provinces ...* (Calcutta: F. Carberry for the Military Orphan Press, 1857) were devoted to listing indigo plantations and factories. The Indigo Planters' Association was established in Calcutta in 1854 to represent the interests of planters.

tation was at a place called Muttra,[1] near the border of the North-west Provinces. Night after night the whole sky was alight with the burning bungalows, and day after day we had small companies of Europeans passing through our estate with their wives and children, on their way to Agra, where were the nearest troops. Mr. Abel White was an obstinate man. He had it in his head that the affair had been exaggerated, and that it would blow over as suddenly as it had sprung up. There he sat on his veranda, drinking whisky-pegs and smoking cheroots,[2] while the country was in a blaze about him. Of course we stuck by him, I and Dawson, who, with his wife, used to do the book-work and the managing. Well, one fine day the crash came. I had been away on a distant plantation and was riding slowly home in the evening, when my eye fell upon something all huddled together at the bottom of a steep nullah.[3] I rode down to see what it was, and the cold struck through my heart when I found it was Dawson's wife, all cut into ribbons, and half eaten by jackals and native dogs. A little further up the road Dawson himself was lying on his face, quite dead, with an empty revolver in his hand, and four sepoys lying across each other in front of him. I reined up my horse, wondering which way I should turn; but at that moment I saw thick smoke curling up from Abel White's bungalow and the flames beginning to burst through the roof. I knew then that I could do my employer no good, but would only throw my own life away if I meddled in the matter. From where I stood I could see hundreds of the black fiends, with their red coats still on their backs, dancing and howling round the burning house. Some of them pointed at me, and a couple of bullets sang past my head: so I broke away across the paddy-fields, and found myself late at night safe within the walls at Agra.

"As it proved, however, there was no great safety there, either. The whole country was up like a swarm of bees. Wherever the English could collect in little bands they held just the ground that their guns commanded. Everywhere else they were helpless fugitives. It was a fight of the millions against the hundreds; and the cruellest part of it was that these men that we fought against,

---

1 Present-day Mathura, a town on the Jamuna river some 50 kilometres (30 miles) northwest of Agra and 150 kilometres (90 miles) south of Delhi, and near the border of what was then the North-West Provinces and Rajputana; today it is in Uttar Pradesh.
2 Cigar of local manufacture, with both ends cut off.
3 Hindi, small stream, gully, or dried up riverbed.

foot, horse, and gunners, were our own picked troops, whom we had taught and trained, handling our own weapons and blowing our own bugle-calls. At Agra there were the 3rd Bengal Fusiliers,[1] some Sikhs, two troops of horse, and a battery of artillery. A volunteer corps of clerks and merchants had been formed, and this I joined, wooden leg and all. We went out to meet the rebels at Shahgunge[2] early in July, and we beat them back for a time, but our powder gave out, and we had to fall back upon the city.

"Nothing but the worst news came to us from every side— which is not to be wondered at, for if you look at the map you will see that we were right in the heart of it. Lucknow[3] is rather better than a hundred miles to the east, and Cawnpore[4] about as far to the south. From every point on the compass there was nothing but torture and murder and outrage.

"The city of Agra is a great place, swarming with fanatics and fierce devil-worshippers of all sorts. Our handful of men were lost among the narrow, winding streets. Our leader moved across the

---

1  This was a British, rather than an Indian regiment. Kaye and Malleson list the British resources in Agra in the summer of 1857 as "five hundred and sixty-eight men of the 3rd European Regiment; one battery with sixty-nine Europeans, including officers, and fifty-four native drivers; fifty-nine mounted militia; and fifty officers and civilians who had taken refuge in Agra." *History of the Indian Mutiny of 1857-8*, III, 180-01.

2  Also spelled Shahgunj, a settlement just outside Agra and today a suburb of the city. The British show of force took place on 5 July 1857.

3  Capital of the self-governing Indian state of Awadh (Oudh), Lucknow, situated on the banks of the River Gumti, was annexed by the British in February 1856. During the "Mutiny" of 1857-78, it was the site of one of the most remarkable episodes in the conflict, the 87-day siege (and successful relief) of the British population (both military and civilian) holed up inside the Lucknow Residency.

4  Modern day Kanpur, a large garrison town on the right bank of the Ganges, half way between Agra and Allahabad. One of the most infamous incidents of the Mutiny took place there on 15 July 1857, when some 200 unarmed British women and children were massacred and their bodies thrown into a well by the forces of the putative Maratha Peshwa, Dundu Panth (c. 1820-59), popularly known as the Nana Sahib.
    Small's geography is decidedly faulty: both Lucknow and Kanpur are to the east rather than to the south of Agra, and they are much further away than Small suggests; Lucknow is 290 kilometres (180 miles) and Kanpur 400 kilometres (250 miles) from Agra.

river, therefore, and took up his position in the old fort[1] of Agra. I don't know if any of you gentlemen have ever read or heard anything of that old fort. It is a very queer place—the queerest that ever I was in, and I have been in some rum corners, too. First of all it is enormous in size. I should think that the enclosure must be acres and acres. There is a modern part, which took all our garrison, women, children, stores, and everything else, with plenty of room over. But the modern part is nothing like the size of the old quarter, where nobody goes, and which is given over to the scorpions and the centipedes. It is all full of great deserted halls, and winding passages, and long corridors twisting in and out, so that it is easy enough for folk to get lost in it. For this reason it was seldom that anyone went into it, though now and again a party with torches might go exploring.

"The river washes along the front of the old fort, and so protects it, but on the sides and behind there are many doors, and these had to be guarded, of course, in the old quarter as well as in that which was actually held by our troops. We were shorthanded, with hardly men enough to man the angles of the building and to serve the guns. It was impossible for us, therefore, to station a strong guard at every one of the innumerable gates. What we did was to organize a central guardhouse in the middle of the fort, and to leave each gate under the charge of one white man and two or three natives. I was selected to take charge during certain hours of the night of a small isolated door upon the south-west side of the building. Two Sikh troopers were placed under my command, and I was instructed if anything went wrong to fire my musket, when I might rely upon help coming at once from the central guard. As the guard was a good two hundred paces away, however, and as the space between was cut up into a labyrinth of passages and corridors, I had great doubts as to whether they could arrive in time to be of any use in case of an actual attack.

"Well, I was pretty proud at having this small command given me, since I was a raw recruit, and a game-legged one at that. For two nights I kept the watch with my Punjaubees.[2] They were tall,

---

1 The current fort, more accurately described as a walled palace, was largely constructed under the Mughal Emperor Akbar (1542-1605), between 1565 and 1573, on the foundations of a smaller former Rajput Fort.

2 Correctly, Punjabis, inhabitants of the Punjab, now divided between India and Pakistan.

fierce-looking chaps, Mahomet Singh and Abdullah Khan by name, both old fighting men, who had borne arms against us at Chilian Wallah.[1] They could talk English pretty well, but I could get little out of them. They preferred to stand together and jabber all night in their queer Sikh lingo. For myself, I used to stand outside the gateway, looking down on the broad, winding river and on the twinkling lights of the great city. The beating of drums, the rattle of tomtoms, and the yells and howls of the rebels, drunk with opium and with bang,[2] were enough to remind us all night of our dangerous neighbours across the stream. Every two hours the officer of the night used to come round to all the posts, to make sure that all was well.

"The third night of my watch was dark and dirty, with a small driving rain. It was dreary work standing in the gateway hour after hour in such weather. I tried again and again to make my Sikhs talk, but without much success. At two in the morning the rounds passed, and broke for a moment the weariness of the night. Finding that my companions would not be led into conversation, I took out my pipe and laid down my musket to strike the match. In an instant the two Sikhs were upon me. One of them snatched my firelock up and levelled it at my head, while the other held a great knife to my throat and swore between his teeth that he would plunge it into me if I moved a step.

"My first thought was that these fellows were in league with the rebels, and that this was the beginning of an assault. If our door were in the hands of the Sepoys the place must fall, and the women and children be treated as they were in Cawnpore.[3] Maybe you gentlemen think that I am just making out a case for myself, but I give you my word that when I thought of that, though I felt the point of the knife at my throat, I opened my mouth with the intention of giving a scream, if it was my last one, which might alarm the main guard. The man who held me seemed to know my thoughts; for, even as I braced myself to it,

1  Decisive battle in the second Anglo-Sikh War that took place between the British and Sikh armies on 13 January 1849, at Chillianwallah, and marked the crushing of independent Sikh power in the Punjab. The British victory was followed by the annexation of the entire province.
2  Also spelled bhang, the narcotic marijuana (*Cannabis sativa*). Often consumed as a drink, sometimes blended with milk, yogurt, or opium suspension, bhang is still used in some Hindu religious rituals.
3  The massacre of women and children in Cawnpore (Kanpur) took place on 15 July 1857; because of the difficulties of communication, it took at least two weeks for news of the massacre to reach Agra.

he whispered: 'Don't make a noise. The fort is safe enough. There are no rebel dogs on this side of the river.' There was the ring of truth in what he said, and I knew that if I raised my voice I was a dead man. I could read it in the fellow's brown eyes. I waited, therefore, in silence, to see what it was that they wanted from me.

"'Listen to me, Sahib,' said the taller and fiercer of the pair, the one whom they called Abdullah Khan. 'You must either be with us now, or you must be silenced forever. The thing is too great a one for us to hesitate. Either you are heart and soul with us on your oath on the cross of the Christians, or your body this night shall be thrown into the ditch, and we shall pass over to our brothers in the rebel army. There is no middle way. Which is it to be—death or life? We can only give you three minutes to decide, for the time is passing, and all must be done before the rounds come again.'

"'How can I decide?' said I. You have not told me what you want of me. But I tell you now that if it is anything against the safety of the fort I will have no truck with it, so you can drive home your knife and welcome.'

"'It is nothing against the fort,' said he. 'We only ask you to do that which your countrymen come to this land for. We ask you to be rich. If you will be one of us this night, we will swear to you upon the naked knife, and by the threefold oath which no Sikh was ever known to break,[1] that you shall have your fair share of the loot. A quarter of the treasure shall be yours. We can say no fairer.'

"'But what is the treasure then?' I asked. 'I am as ready to be rich as you can be if you will but show me how it can be done.'

"'You will swear, then,' said he, 'by the bones of your father, by the honour of your mother, by the cross of your faith, to raise no hand and speak no word against us, either now or afterwards?'

"'I will swear it,' I answered, 'provided that the fort is not endangered.'

"'Then my comrade and I will swear that you shall have a quarter of the treasure which shall be equally divided among the four of us.'

"'There are but three,' said I.

"'No; Dost Akbar must have his share. We can tell the tale to you while we wait them. Do you stand at the gate, Mahomet

---

1   Doyle's garbled and somewhat sacrilegious borrowing is from the Sikh
    initiation into the Khalsa (or community of observant Sikhs), the Amrit
    Sanskar ceremony, during which each person vows to uphold the five
    (*not* three) principles of Sikhism.

Singh, and give notice of their coming. The thing stands thus, Sahib, and I tell it to you because I know that an oath is binding upon a Feringhee,[1] and that we may trust you. Had you been a lying Hindoo, though you had sworn by all the gods in their false temples, your blood would have been upon the knife and your body in the water. But the Sikh knows the Englishman, and the Englishman knows the Sikh.[2] Hearken, then, to what I have to say.

"'There is a rajah in the northern provinces who has much wealth, though his lands are small. Much has come to him from his father, and more still he has set by himself, for he is of a low nature, and hoards his gold rather than spend it. When the troubles broke out he would be friends both with the lion and the tiger—with the Sepoy and with the Company's Raj. Soon, however, it seemed to him that the white men's day was come, for through all the land he could hear of nothing but of their death and their overthrow. Yet, being a careful man, he made such plans that, come what might, half at least of his treasure should be left to him. That which was in gold and silver he kept by him in the vaults of his palace, but the most precious stones and the choicest pearls that he had he put in an iron box and sent it by a trusty servant, who, under the guise of a merchant, should take it to the fort at Agra, there to lie until the land is at peace. Thus, if the rebels won he would have his money, but if the Company conquered, his jewels would be saved to him. Having thus divided his hoard, he threw himself into the cause of the Sepoys, since they were strong upon his borders. By his doing this, mark you, Sahib, his property becomes the due of those who have been true to their salt.

"'This pretended merchant, who travels under the name of Achmet, is now in the city of Agra, and desires to gain his way into the fort. He has with him as travelling-companion my foster-brother Dost Akbar, who knows his secret. Dost Akbar has promised this night to lead him to a side-postern of the fort, and has

---

1 Hindi, foreigner. Originally specifically used for the French, but, by this time, a generically derogatory term for all Europeans.

2 A relatively recent development. The Sikhs had fought the British twice, in the First (1845-46) and the Second (1848-49) Anglo-Sikh Wars, but at the time of the Indian "Mutiny," the British had actively recruited large numbers of Sikhs into military service. Classified as a "martial race," this policy of Sikh recruitment would continue until Indian independence.

chosen this one for his purpose. Here he will come presently, and here he will find Mahomet Singh and myself awaiting him. The place is lonely, and none shall know of his coming. The world shall know the merchant Achmet no more, but the great treasure of the rajah shall be divided among us. What say you to it, Sahib?'

"In Worcestershire the life of a man seems a great and a sacred thing; but it is very different when there is fire and blood all round you, and you have been used to meeting death at every turn. Whether Achmet the merchant lived or died was a thing as light as air to me, but at the talk about the treasure my heart turned to it, and I thought of what I might do in the old country with it, and how my folk would stare when they saw their ne'er-do-well coming back with his pockets full of gold moidores.[1] I had, therefore, already made up my mind. Abdullah Khan, however, thinking that I hesitated, pressed the matter more closely.

"'Consider, Sahib,' said he, 'that if this man is taken by the commandant he will be hung or shot, and his jewels taken by the government, so that no man will be a rupee the better for them. Now, since we do the taking of him, why should we not do the rest as well? The jewels will be as well with us as in the Company's coffers. There will be enough to make every one of us rich men and great chiefs. No one can know about the matter, for here we are cut off from all men. What could be better for the purpose? Say again, then, Sahib, whether you are with us, or if we must look upon you as an enemy.'

"'I am with you heart and soul,' said I.

"'It is well,' he answered, handing me back my firelock. 'You see that we trust you, for your word, like ours, is not to be broken. We have now only to wait for my brother and the merchant.'

"'Does your brother know, then, of what you will do?' I asked.

"'The plan is his. He has devised it. We will go to the gate and share the watch with Mahomet Singh.'

"The rain was still falling steadily, for it was just the beginning of the wet season. Brown, heavy clouds were drifting across the sky, and it was hard to see more than a stonecast. A deep moat lay in front of our door, but the water was in places nearly dried up, and it could easily be crossed. It was strange to me to be standing there with those two wild Punjaubees waiting for the

---

1   From the Portuguese, moeda d'ouro (coin of gold), a gold coin of Portuguese origin in common use in England and its colonies in the first half of the eighteenth century. It was worth 27 shillings.

man who was coming to his death.

"Suddenly my eye caught the glint of a shaded lantern at the other side of the moat. It vanished among the mound-heaps, and then appeared again coming slowly in our direction.

"'Here they are!' I exclaimed.

"'You will challenge him, Sahib, as usual,' whispered Abdullah. 'Give him no cause for fear. Send us in with him, and we shall do the rest while you stay here on guard. Have the lantern ready to uncover, that we may be sure that it is indeed the man.'

"The light had flickered onwards, now stopping and now advancing, until I could see two dark figures upon the other side of the moat. I let them scramble down the sloping bank, splash through the mire, and climb halfway up to the gate before I challenged them.

"'Who goes there?' said I in a subdued voice.

"'Friends,' came the answer. I uncovered my lantern and threw a flood of light upon them. The first was an enormous Sikh, with a black beard which swept nearly down to his cummerbund. Outside of a show I have never seen so tall a man. The other was a little fat, round fellow, with a great yellow turban and a bundle in his hand, done up in a shawl. He seemed to be all in a quiver with fear, for his hands twitched as if he had the ague, and his head kept turning to left and right with two bright little twinkling eyes, like a mouse when he ventures out from his hole. It gave me the chills to think of killing him, but I thought of the treasure, and my heart set as hard as a flint within me. When he saw my white face he gave a little chirrup of joy and came running up towards me.

"'Your protection, sahib,' he panted, 'your protection for the unhappy merchant Achmet. I have travelled across Rajpootana, that I might seek the shelter of the fort at Agra. I have been robbed and beaten and abused because I have been the friend of the Company. It is a blessed night this when I am once more in safety—I and my poor possessions.'

"'What have you in the bundle?' I asked.

"'An iron box,' he answered, 'which contains one or two little family matters which are of no value to others but which I should be sorry to lose. Yet I am not a beggar; and I shall reward you, young Sahib, and your governor also if he will give me the shelter I ask.'

"I could not trust myself to speak longer with the man. The more I looked at his fat, frightened face, the harder did it seem that we should slay him in cold blood. It was best to get it over.

"'Take him to the main guard,' said I. The two Sikhs closed in upon him on each side, and the giant walked behind, while they marched in through the dark gateway. Never was a man so compassed round with death. I remained at the gateway with the lantern.

"I could hear the measured tramp of their footsteps sounding through the lonely corridors. Suddenly it ceased, and I heard voices and a scuffle, with the sound of blows. A moment later there came, to my horror, a rush of footsteps coming in my direction, with a loud breathing of a running man. I turned my lantern down the long straight passage, and there was the fat man, running like the wind, with a smear of blood across his face, and close at his heels, bounding like a tiger, the great black-bearded Sikh, with a knife flashing in his hand. I have never seen a man run so fast as that little merchant. He was gaining on the Sikh, and I could see that if he once passed me and got to the open air he would save himself yet. My heart softened to him, but again the thought of his treasure turned me hard and bitter. I cast my firelock between his legs as he raced past, and he rolled twice over like a shot rabbit. Ere he could stagger to his feet the Sikh was upon him and buried his knife twice in his side. The man never uttered moan nor moved muscle, but lay where he had fallen. I think myself that he may have broken his neck with the fall. You see, gentlemen, that I am keeping my promise. I am telling you every word of the business just exactly as it happened, whether it is in my favour or not."

He stopped and held out his manacled hands for the whisky-and-water which Holmes had brewed for him. For myself, I confess that I had now conceived the utmost horror of the man not only for this cold-blooded business in which he had been concerned but even more for the somewhat flippant and careless way in which he narrated it. Whatever punishment was in store for him, I felt that he might expect no sympathy from me. Sherlock Holmes and Jones sat with their hands upon their knees, deeply interested in the story but with the same disgust written upon their faces. He may have observed it, for there was a touch of defiance in his voice and manner as he proceeded.

"It was all very bad, no doubt," said he. "I should like to know how many fellows in my shoes would have refused a share of this loot when they knew that they would have their throats cut for their pains. Besides, it was my life or his when once he was in the fort. If he had got out, the whole business would come to light, and I should have been court-martialled and shot as likely as not; for people were not very lenient at a time like that."

"Go on with your story," said Holmes shortly.

"Well, we carried him in, Abdullah, Akbar, and I. A fine weight he was, too, for all that he was so short. Mahomet Singh was left to guard the door. We took him to a place which the Sikhs had already prepared. It was some distance off, where a winding passage leads to a great empty hall, the brick walls of which were all crumbling to pieces. The earth floor had sunk in at one place, making a natural grave, so we left Achmet the merchant there, having first covered him over with loose bricks. This done, we all went back to the treasure.

"It lay where he had dropped it when he was first attacked. The box was the same which now lies open upon your table. A key was hung by a silken cord to that carved handle upon the top. We opened it, and the light of the lantern gleamed upon a collection of gems such as I have read of and thought about when I was a little lad at Pershore. It was blinding to look upon them. When we had feasted our eyes we took them all out and made a list of them. There were one hundred and forty-three diamonds of the first water, including one which has been called, I believe, 'the Great Mogul,'[1] and is said to be the second largest stone in existence. Then there were ninety-seven very fine emeralds, and one hundred and seventy rubies, some of which, however, were small. There were forty carbuncles, two hundred and ten sapphires, sixty-one agates, and a great quantity of beryls, onyxes, cats'-eyes, turquoises, and other stones, the very names of which I did not know at the time, though I have become more familiar with them since. Besides this, there were nearly three hundred very fine pearls, twelve of which were set in a gold coronet. By the way, these last had been taken out of the chest, and were not there when I recovered it.

---

1  The name given to the largest diamond ever found in India, this was a 787 carat rough stone found in the Kollur mine in Golconda in 1650. The first European account of the fabled stone was by François Bernier, who saw it at the court of the Mughal Emperor Shah Jahan in Agra in 1655, and it was again noted by Jean-Baptiste Tavernier in 1665. There are no further accounts of the Great Mogul diamond after Nadir Shah's sacking of Delhi in 1738-39, and the stone may well have been cut up and dispersed after that. Speculation about the fate of the Great Mogul was still rife in the nineteenth century. Until the early eighteenth century, India's Golconda mines were the world's sole source for diamonds.

"After we had counted our treasures we put them back into the chest and carried them to the gateway to show them to Mahomet Singh. Then we solemnly renewed our oath to stand by each other and be true to our secret. We agreed to conceal our loot in a safe place until the country should be at peace again, and then to divide it equally among ourselves. There was no use dividing it at present, for if gems of such value were found upon us it would cause suspicion, and there was no privacy in the fort nor any place where we could keep them. We carried the box, therefore, into the same hall where we had buried the body, and there, under certain bricks in the best-preserved wall, we made a hollow and put our treasure. We made careful note of the place, and next day I drew four plans, one for each of us, and put the sign of the four of us at the bottom, for we had sworn that we should each always act for all, so that none might take advantage. That is an oath that I can put my hand to my heart and swear that I have never broken.

"Well, there's no use my telling you gentlemen what came of the Indian mutiny. After Wilson[1] took Delhi and Sir Colin[2] relieved Lucknow the back of the business was broken. Fresh troops came pouring in, and Nana Sahib made himself scarce over the frontier. A flying column under Colonel Greathed[3] came round to Agra and cleared the Pandies away from it. Peace seemed to be settling upon the country, and we four were beginning to hope that the time was at hand when we might safely go off with our shares of the plunder. In a moment, however, our hopes were shattered by our being arrested as the murderers of Achmet.

"It came about in this way. When the rajah put his jewels into the hands of Achmet he did it because he knew that he was a trusty man. They are suspicious folk in the East, however: so what does this rajah do but take a second even more trusty servant and

---

1  Sir Archdale Wilson (1803-74), Brigadier of the Artillery and previously commander of the army at Meerut (where the "Mutiny" had started), led the assault to successfully recapture Delhi, 13-20 September 1857. He was knighted on 17 November 1857.

2  Sir Colin Campbell (1792-1863), commander of the expeditionary force sent to India on 12 July 1857, planned and executed the relief of the besieged British in the Residency in Lucknow. The 87-day siege was lifted on 17 November 1857.

3  Lieutenant-Colonel Edward Greathed (1812-81), who led the army that recaptured Agra on 10 October 1857, made use of considerable numbers of new Sikh recruits.

set him to play the spy upon the first? This second man was ordered never to let Achmet out of his sight, and he followed him like his shadow. He went after him that night and saw him pass through the doorway. Of course he thought he had taken refuge in the fort and applied for admission there himself next day, but could find no trace of Achmet. This seemed to him so strange that he spoke about it to a sergeant of guides, who brought it to the ears of the commandant. A thorough search was quickly made, and the body was discovered. Thus at the very moment that we thought that all was safe we were all four seized and brought to trial on a charge of murder—three of us because we had held the gate that night, and the fourth because he was known to have been in the company of the murdered man. Not a word about the jewels came out at the trial, for the rajah had been deposed and driven out of India: so no one had any particular interest in them. The murder, however, was clearly made out, and it was certain that we must all have been concerned in it. The three Sikhs got penal servitude for life, and I was condemned to death,[1] though my sentence was afterwards commuted to the same as the others.

"It was rather a queer position that we found ourselves in then. There we were all four tied by the leg and with precious little chance of ever getting out again, while we each held a secret which might have put each of us in a palace if we could only have made use of it. It was enough to make a man eat his heart out to have to stand the kick and the cuff of every petty jack-in-office, to have rice to eat and water to drink, when that gorgeous fortune was ready for him outside, just waiting to be picked up. It might have driven me mad; but I was always a pretty stubborn one, so I just held on and bided my time.

"At last it seemed to me to have come. I was changed from Agra to Madras, and from there to Blair Island[2] in the Andamans. There are very few white convicts at this settlement, and, as I had behaved well from the first, I soon found myself a sort of privileged person. I was given a hut in Hope Town, which is a small place on the slopes of Mount Harriet,[3] and I was left pretty much to myself. It is a dreary, fever-stricken place, and all

---

1  Once again, the difference in the sentences suggests that Small is being economical with the truth and his own culpability.
2  Correctly, Port Blair on Great Andaman, the largest of the Andaman Islands.
3  See Appendix D3.

beyond our little clearings was infested with wild cannibal natives, who were ready enough to blow a poisoned dart at us if they saw a chance. There was digging and ditching and yam-planting, and a dozen other things to be done, so we were busy enough all day; though in the evening we had a little time to ourselves. Among other things, I learned to dispense drugs for the surgeon, and picked up a smattering of his knowledge. All the time I was on the look-out for a chance to escape; but it is hundreds of miles from any other land, and there is little or no wind in those seas: so it was a terribly difficult job to get away.

"The surgeon, Dr. Somerton, was a fast, sporting young chap, and the other young officers would meet in his rooms of an evening and play cards. The surgery, where I used to make up my drugs, was next to his sitting-room, with a small window between us. Often, if I felt lonesome, I used to turn out the lamp in the surgery, and then, standing there, I could hear their talk and watch their play. I am fond of a hand at cards myself, and it was almost as good as having one to watch the others. There was Major Sholto, Captain Morstan, and Lieutenant Bromley Brown, who were in command of the native troops, and there was the surgeon himself, and two or three prison-officials, crafty old hands who played a nice sly safe game. A very snug little party they used to make.

"Well, there was one thing which very soon struck me, and that was that the soldiers used always to lose and the civilians to win. Mind, I don't say there was anything unfair, but so it was. These prison-chaps had done little else than play cards ever since they had been at the Andamans, and they knew each other's game to a point, while the others just played to pass the time and threw their cards down anyhow. Night after night the soldiers got up poorer men, and the poorer they got the more keen they were to play. Major Sholto was the hardest hit. He used to pay in notes and gold at first, but soon it came to notes of hand and for big sums. He sometimes would win for a few deals, just to give him heart, and then the luck would set in against him worse than ever. All day he would wander about as black as thunder, and he took to drinking a deal more than was good for him.

"One night he lost even more heavily than usual. I was sitting in my hut when he and Captain Morstan came stumbling along on the way to their quarters. They were bosom friends, those two, and never far apart. The major was raving about his losses.

"'It's all up, Morstan,' he was saying as they passed my hut. 'I shall have to send in my papers. I am a ruined man.'

"'Nonsense, old chap!' said the other, slapping him upon the shoulder. 'I've had a nasty facer myself, but—' That was all I could hear, but it was enough to set me thinking.

"A couple of days later Major Sholto was strolling on the beach: so I took the chance of speaking to him.

"'I wish to have your advice, Major,' said I.

"'Well, Small, what is it?' he asked, taking his cheroot from his lips.

"'I wanted to ask you, sir,' said I, 'who is the proper person to whom hidden treasure should be handed over. I know where half a million worth lies, and, as I cannot use it myself, I thought perhaps the best thing that I could do would be to hand it over to the proper authorities, and then perhaps they would get my sentence shortened for me.'

"'Half a million, Small?' he gasped, looking hard at me to see if I was in earnest.

"'Quite that, sir—in jewels and pearls. It lies there ready for anyone. And the queer thing about it is that the real owner is outlawed and cannot hold property, so that it belongs to the first comer.'

"'To Government, Small,' he stammered, 'to Government.' But he said it in a halting fashion, and I knew in my heart that I had got him.

"'You think, then, sir, that I should give the information to the Governor-General?' said I quietly.

"'Well, well, you must not do anything rash, or that you might repent. Let me hear all about it, Small. Give me the facts.'

"I told him the whole story, with small changes, so that he could not identify the places. When I had finished he stood stock still and full of thought. I could see by the twitch of his lip that there was a struggle going on within him.

"'This is a very important matter, Small,' he said at last. 'You must not say a word to anyone about it, and I shall see you again soon.'

"Two nights later he and his friend, Captain Morstan, came to my hut in the dead of the night with a lantern.

"'I want you just to let Captain Morstan hear that story from your own lips, Small,' said he.

"I repeated it as I had told it before.

"'It rings true, eh?' said he. 'It's good enough to act upon?'

"Captain Morstan nodded.

"'Look here, Small,' said the Major. 'We have been talking it over, my friend here and I, and we have come to the conclusion that this secret of yours is hardly a Government matter, after all, but is a private concern of your own, which of course you have the power of disposing of as you think best. Now the question is, What price would you ask for it? We might be inclined to take it up, and at least look into it, if we could agree as to terms.' He tried to speak in a cool, careless way, but his eyes were shining with excitement and greed.

"'Why, as to that, gentlemen,' I answered, trying also to be cool, but feeling as excited as he did, 'there is only one bargain which a man in my position can make. I shall want you to help me to my freedom, and to help my three companions to theirs. We shall then take you into partnership and give you a fifth share to divide between you.'

"'Hum!' said he. 'A fifth share! That is not very tempting.'

"'It would come to fifty thousand apiece,' said I.

"'But how can we gain your freedom? You know very well that you ask an impossibility.'

"'Nothing of the sort,' I answered. 'I have thought it all out to the last detail. The only bar to our escape is that we can get no boat fit for the voyage, and no provisions to last us for so long a time. There are plenty of little yachts and yawls[1] at Calcutta or Madras which would serve our turn well. Do you bring one over. We shall engage to get aboard her by night, and if you will drop us on any part of the Indian coast you will have done your part of the bargain.'

"'If there were only one,' he said.

"'None or all,' I answered. 'We have sworn it. The four of us must always act together.'

"'You see, Morstan,' said he, 'Small is a man of his word. He does not flinch from his friends. I think we may very well trust him.'

"'It's a dirty business,' the other answered. 'Yet, as you say, the money will save our commissions handsomely.'

"'Well, Small,' said the Major, 'we must, I suppose, try and meet you. We must first, of course, test the truth of your story. Tell me where the box is hid, and I shall get leave of absence and go back to India in the monthly relief-boat to inquire into the affair.'

"'Not so fast,' said I, growing colder as he got hot. 'I must have the consent of my three comrades. I tell you that it is four or none with us.'

---

1  Small sailing boat.

"'Nonsense!' he broke in. 'What have three black fellows to do with our agreement?'

"'Black or blue,' said I, 'they are in with me, and we all go together.'

"Well, the matter ended by a second meeting, at which Mahomet Singh, Abdullah Khan, and Dost Akbar were all present. We talked the matter over again, and at last we came to an arrangement. We were to provide both the officers with charts of the part of the Agra fort, and mark the place in the wall where the treasure was hid. Major Sholto was to go to India to test our story. If he found the box he was to leave it there, to send out a small yacht provisioned for a voyage, which was to lie off Rutland Island,[1] and to which we were to make our way, and finally to return to his duties. Captain Morstan was then to apply for leave of absence, to meet us at Agra, and there we were to have a final division of the treasure, he taking the major's share as well as his own. All this we sealed by the most solemn oaths that the mind could think or the lips utter. I sat up all night with paper and ink, and by the morning I had the two charts all ready, signed with the sign of four—that is, of Abdullah, Akbar, Mahomet, and myself.

"Well, gentlemen, I weary you with my long story, and I know that my friend Mr. Jones is impatient to get me safely stowed in chokey.[2] I'll make it as short as I can. The villain Sholto went off to India, but he never came back again. Captain Morstan showed me his name among a list of passengers in one of the mail-boats very shortly afterwards. His uncle had died, leaving him a fortune, and he had left the army; yet he could stoop to treat five men as he had treated us. Morstan went over to Agra shortly afterwards and found, as we expected, that the treasure was indeed gone. The scoundrel had stolen it all, without carrying out one of the conditions on which we had sold him the secret. From that day I lived only for vengeance. I thought of it by day and I nursed it by night. It became an overpowering, absorbing passion with me. I cared nothing for the law—nothing for the gallows. To escape, to track down Sholto, to have my hand upon his throat—that was my one thought. Even the Agra treasure had come to be a smaller thing in my mind than the slaying of Sholto.

"Well, I have set my mind on many things in this life, and never one which I did not carry out. But it was weary years before

---

1  Rutland Island is immediately to the south of Great Andaman, across the MacPherson Strait.
2  Anglo-Indian slang word for prison.

my time came. I have told you that I had picked up something of medicine. One day when Dr. Somerton was down with a fever a little Andaman Islander was picked up by a convict-gang in the woods. He was sick to death, and had gone to a lonely place to die. I took him in hand, though he was as venomous as a young snake, and after a couple of months I got him all right and able to walk. He took a kind of fancy to me then, and would hardly go back to his woods, but was always hanging about my hut. I learned a little of his lingo from him, and this made him all the fonder of me.

"Tonga—for that was his name—was a fine boatman and owned a big, roomy canoe of his own. When I found that he was devoted to me and would do anything to serve me, I saw my chance of escape. I talked it over with him. He was to bring his boat round on a certain night to an old wharf which was never guarded, and there he was to pick me up. I gave him directions to have several gourds of water and a lot of yams, cocoanuts, and sweet potatoes.

"He was staunch and true, was little Tonga. No man ever had a more faithful mate. At the night named he had his boat at the wharf. As it chanced, however, there was one of the convict-guard down there—a vile Pathan[1] who had never missed a chance of insulting and injuring me. I had always vowed vengeance, and now I had my chance. It was as if fate had placed him in my way that I might pay my debt before I left the island. He stood on the bank with his back to me, and his carbine on his shoulder. I looked about for a stone to beat out his brains with, but none could I see.

"Then a queer thought came into my head and showed me where I could lay my hand on a weapon. I sat down in the darkness and unstrapped my wooden leg. With three long hops I was on him. He put his carbine to his shoulder, but I struck him full, and knocked the whole front of his skull in. You can see the split in the wood now where I hit him. We both went down together, for I could not keep my balance; but when I got up I found him still lying quiet enough. I made for the boat, and in an hour we

---

1  The Pathans are the predominant ethnic group in Afghanistan and in the Northwest Frontier Province of present-day Pakistan. Classified as a "martial race" after 1857, they were often recruited by the British as soldiers and policemen, despite the background of mutual animosity caused by the continuing Anglo-Afghan Wars. Lord Mayo's assassin was a Pathan; see Appendix C4.

were well out at sea. Tonga had brought all his earthly possessions with him, his arms and his gods.[1] Among other things, he had a long bamboo spear, and some Andaman cocoa-nut matting,[2] with which I made a sort of a sail. For ten days we were beating about, trusting to luck, and on the eleventh we were picked up by a trader which was going from Singapore to Jiddah with a cargo of Malay pilgrims.[3] They were a rum crowd, and Tonga and I soon managed to settle down among them. They had one very good quality: they let you alone and asked no questions.

"Well, if I were to tell you all the adventures that my little chum and I went through, you would not thank me, for I would have you here until the sun was shining. Here and there we drifted about the world, something always turning up to keep us from London. All the time, however, I never lost sight of my purpose. I would dream of Sholto at night. A hundred times I have killed him in my sleep. At last, however, some three or four years ago,[4] we found ourselves in England. I had no great difficulty in finding where Sholto lived, and I set to work to discover whether he had realized on the treasure, or if he still had it. I made friends with someone who could help me—I name no names, for I don't want to get anyone else in a hole—and I soon found that he still had the jewels. Then I tried to get at him in many ways; but he was pretty sly and had always two prize-fighters, besides his sons and his khitmutgar,[5] on guard over him.

"One day, however, I got word that he was dying. I hurried at once to the garden, mad that he should slip out of my clutches like that, and, looking through the window, I saw him lying in his bed, with his sons on each side of him. I'd have come through and

---

1  In fact, the Andamanese did not possess images of gods to worship.

2  This would have to have been a traded item, for until the arrival of the British, there were no coconut palms (and hence no coconut fibre matting) on the Andaman Islands.

3  A common enough journey, and one that would later become memorialised in British fiction in Joseph Conrad's *Lord Jim* (1900).

4  Another inconsistency in Small's narrative, for he states that he returned to England in 1884-85, and Thaddeus Sholto tells us (and Holmes confirms by examining the obituaries in *The Times*) that Major John Sholto dies, suffering from an enlarged spleen, on 28 April 1882.

5  Another potential inconsistency, for Lal Chowdar was already dead (he pre-deceased Major Sholto), and his presumed successor, Lal Rao, is confirmed by Holmes as being Small's confederate inside the house—so why would he worry about the apparent obstacle presented by the khitmutgar?

taken my chance with the three of them, only even as I looked at him his jaw dropped, and I knew that he was gone. I got into his room that same night, though, and I searched his papers to see if there was any record of where he had hidden our jewels. There was not a line, however, so I came away, bitter and savage as a man could be. Before I left I bethought me that if I ever met my Sikh friends again it would be a satisfaction to know that I had left some mark of our hatred; so I scrawled down the sign of the four of us, as it had been on the chart, and I pinned it on his bosom. It was too much that he should be taken to the grave without some token from the men whom he had robbed and befooled.

"We earned a living at this time by my exhibiting poor Tonga at fairs and other such places as the black cannibal. He would eat raw meat and dance his war-dance: so we always had a hatful of pennies after a day's work. I still heard all the news from Pondicherry Lodge, and for some years there was no news to hear, except that they were hunting for the treasure. At last, however, came what we had waited for so long. The treasure had been found. It was up at the top of the house in Mr. Bartholomew Sholto's chemical laboratory. I came at once and had a look at the place, but I could not see how, with my wooden leg, I was to make my way up to it. I learned, however, about a trapdoor in the roof, and also about Mr. Sholto's supper-hour. It seemed to me that I could manage the thing easily through Tonga. I brought him out with me with a long rope wound round his waist. He could climb like a cat, and he soon made his way through the roof, but, as ill luck would have it, Bartholomew Sholto was still in the room, to his cost. Tonga thought he had done something very clever in killing him, for when I came up by the rope I found him strutting about as proud as a peacock. Very much surprised was he when I made at him with the rope's end and cursed him for a little bloodthirsty imp. I took the treasure box and let it down, and then slid down myself, having first left the sign of the four upon the table to show that the jewels had come back at last to those who had most right to them. Tonga then pulled up the rope, closed the window, and made off the way that he had come.

"I don't know that I have anything else to tell you. I had heard a waterman speak of the speed of Smith's launch, the *Aurora*, so I thought she would be a handy craft for our escape. I engaged with old Smith, and was to give him a big sum if he got us safe to our ship. He knew, no doubt, that there was some screw loose, but he was not in our secrets. All this is the truth, and if I tell it to you, gentlemen, it is not to amuse you—for you have not done me a very good turn—but it is because I believe the best defence I can make is just to hold back nothing, but let all the world know how badly I have myself been served by Major Sholto, and how innocent I am of the death of his son."

"A very remarkable account," said Sherlock Holmes. "A fitting wind-up to an extremely interesting case. There is nothing at all new to me in the latter part of your narrative except that you brought your own rope. That I did not know. By the way, I had hoped that Tonga had lost all his darts; yet he managed to shoot one at us in the boat."

"He had lost them all, sir, except the one which was in his blow-pipe at the time."

"Ah, of course," said Holmes. "I had not thought of that."

"Is there any other point which you would like to ask about?" asked the convict affably.

"I think not, thank you," my companion answered.

"Well, Holmes," said Athelney Jones, "you are a man to be humoured, and we all know that you are a connoisseur of crime; but duty is duty, and I have gone rather far in doing what you and your friend asked me. I shall feel more at ease when we have our story-teller here safe under lock and key. The cab still waits, and there are two inspectors downstairs. I am much obliged to you both for your assistance. Of course you will be wanted at the trial. Good-night to you."

"Good-night, gentlemen both," said Jonathan Small.

"You first, Small," remarked the wary Jones as they left the room. "I'll take particular care that you don't club me with your wooden leg, whatever you may have done to the gentleman at the Andaman Isles."

"Well, and there is the end of our little drama," I remarked after we had sat some time smoking in silence. "I fear that it may be the last investigation in which I shall have the chance of studying your methods. Miss Morstan has done me the honour to accept me as a husband in prospective."

He gave a most dismal groan.

"I feared as much," said he. "I really cannot congratulate you."

I was a little hurt.

"Have you any reason to be dissatisfied with my choice?" I asked.

"Not at all. I think she is one of the most charming young ladies I ever met, and might have been most useful in such work as we have been doing. She had a decided genius that way; witness the way in which she preserved that Agra plan from all the other papers of her father. But love is an emotional thing, and whatever is emotional is opposed to that true cold reason which I place above all things. I should never marry myself, lest I bias my judgment."

"I trust," said I, laughing, "that my judgment may survive the ordeal. But you look weary."

"Yes, the reaction is already upon me. I shall be as limp as a rag for a week."

"Strange," said I, "how terms of what in another man I should call laziness alternate with your fits of splendid energy and vigour."

"Yes," he answered, "there are in me the makings of a very fine loafer, and also of a pretty spry sort of a fellow. I often think of those lines of old Goethe: 'Schade dass die Natur nur *einen* Mensch aus dir schuf, Denn zum würdigen Mann war und zum Schelmen der Stoff.'[1] By the way, *à propos* of this Norwood business, you see that they had, as I surmised, a confederate in the house, who could be none other than Lal Rao, the butler: so Jones actually has the undivided honour of having caught one fish in his great haul."

"The division seems rather unfair," I remarked. "You have done all the work in this business. I get a wife out of it, Jones gets the credit, pray what remains for you?"

"For me," said Sherlock Holmes, "there still remains the cocaine-bottle." And he stretched his long white hand up for it.

## THE END

---

1 German, "unfortunately, nature made only one being out of you, although there was enough material for a good man and a rogue": epigram number 20 ("Der Prophet") from Wolfgang von Goethe and Friedrich von Schiller's dialogue poem, *Xenien* (1797). Grammatically it should be "einen Menschen."

# Appendix A: Domestic Contexts

[*The Sign of Four* is an unashamedly contemporary novel, written rapidly and designed to be consumed quickly by a new, increasingly urban, physically mobile, and technologically sophisticated popular readership. As such, it reflects the development of popular readers and their concerns by engaging with a number of important domestic issues current in British society in the 1880s and 1890s. The first chapter of the book is titled "The Science of Deduction," consciously repeating the title of the second chapter of the very first Sherlock Holmes story, *A Study in Scarlet* (1887). Doyle and his publishers were still in the process of creating Sherlock Holmes for the reading public, and this is evident in the repeated and deliberate enunciation of Holmes's method to his readers. The emphasis on deduction not only demonstrates Holmes's approach, but it also clearly echoed current thinking in a wide range of British scientific and pseudo-scientific fields, from criminology and ethnography to psychology and physiology. Holmes's ability to construct an accurate criminal profile of Jonathan Small from collected data reflects late nineteenth-century work in criminology; the innately criminal man that he fashions, complete with excessive facial hair and prognathism, is straight out of the studies of criminal types prevalent at the time, best exemplified by the work of Cesare Lombroso. *The Sign of Four* is also one of the first great expositions of the character and personality of Sherlock Holmes, and it is certainly worth thinking about the presentation of his fictional genius in the light of contemporaneous work by Ellis, Galton, and Lombroso about actual genius. Social scientists, criminologists, anthropologists, and eugenicists in this period were all committed to the identification and explication of human genius, but at the same time they constantly drew parallels with their work in criminology. Doyle's narrative presents both, and demonstrates the potency and ubiquity of these ideas in late nineteenth-century British culture and society.]

## 1. From Havelock Ellis, *The Criminal* (London: Walter Scott, 1890)

[This book was published by Walter Scott in their Contemporary Science Series, for which Havelock Ellis (1859-1939) was general editor. Ellis was the pre-eminent sexologist of his generation; his multivolume *Studies in the Psychology of Sex* (1897-1910) was widely considered to be the most important examination of the subject in the period. Published in cloth-bound crown octavo volumes, and priced at

3s 6d per title, the series was directly pitched at general readers. Appendix B in *The Criminal* (307-16) gives a detailed account of the Second Congress of Criminal Anthropology, which took place in Paris in August 1889. Jonathan Small's physiognomy (prominent chin, extensively wrinkled face, and abundance of dark hair), fits a widespread pseudoscientific idea of a "criminal type" prevalent at the time.]

*On prognathism*

Prognathism has frequently been noted as a prominent characteristic of the criminal face, both in men and women ... there is little doubt that the lower jaw is often remarkably well developed in those guilty of crimes of violence. The squareness and prominence of the jaw are obvious to the eye, and this is verified by weighing after death ... the average weight of the Parisian criminal skull is, if anything, below that of the ordinary Parisian, but while the average weight of the lower jaw in the latter is about 80 grammes, it is about 94 grammes among murderers. In this respect, the criminal resembles the savage and the prehistoric man; among the insane the jaw weighs rather less than the normal average. (63-64)

*On wrinkles*

Ottolonghi has investigated the wrinkles on the faces of 200 criminals as compared with 200 normal persons. He finds that they are much more frequent and much more marked in the criminal than in the non-criminal person, and this must have struck many persons who have seen a large number of criminals or photographs of criminals. (72)

*On hirsuteness*

The largest proportion of full beards among criminals was found by Marro[1] in sexual offenders ... this abundance of hair seems to be correlated with the animal vigour which is often so noticeable among criminals. It may at the same time be to some extent explained by arrest of development or atavism ... in regard to colours, the proportion of dark-haired persons is considered greater among criminals than among the ordinary population in England, Italy and Germany. (72-74)

---

1  Antonio Marro, social anthropologist and colleague of Cesare Lombroso at Turin University and author of *I caratteri dei delinquenti* (1887).

Of 129 persons "wanted" at Scotland Yard, I find that 45 have "dark brown" hair, and of these 17 (*i.e.*, 37.7 per cent) are described as "dangerous," "desperate," "expert," or "notorious"; 46 have "brown" hair, and of these 14 (*i.e.*, 30 per cent) are "dangerous," etc; 11 are "dark" (9) or "black" (2), and of these 3 (*i.e.*, 27.2 per cent) are "dangerous"; 27 are described as "light brown," "light," "sandy," "fair," "auburn" (one, a woman), "red" (one, a man, who is "dangerous"), and of these 9 (*i.e.*, 33.3 per cent) are "dangerous" etc. This gives a proportion of red-haired persons about the same, according to my observations, as is found among middle-class men in the city, but considerably lower than is found, according to Dr. Beddoe,[1] the chief authority on this subject (in his *Races of Britain*), among the lower classes in London—*i.e.*, about 4 per cent ... so far as exact evidence on the colour of the hair goes, it points chiefly to a relative deficiency of red-haired persons among criminals. (76-77)

## 2. From Cesare Lombroso, *The Man of Genius* (London: Walter Scott, 1891)

[Cesare Lombroso (1835-1909) pioneered the systematic study of criminality; his *L'uomo delinquente* ("Criminal Man") first published in 1876, was one of the most important works of social and criminal pseudoscience in the late nineteenth century. In *L'uomo delinquente*, Lombroso catalogued thousands of photographs of different criminals and grouped them according to specific physiognomic features, thereby creating a number of physically differentiated (and presumably, identifiable) criminal types. Allied with the new technology of photography, Lombroso's work was massively influential, not least in shaping policing and criminology in colonial India.

Published in the same Contemporary Science Series edited by Ellis, Lombroso's *The Man of Genius* attempted to examine genius through mental phenomena. For Lombroso, genius was a "special morbid condition" (v) like any other; in the book, he repeatedly compares genius to insanity. Various attempts to define and quantify genius proliferated in the late nineteenth century; Francis Galton's statistical approach, *Hereditary Genius: an Inquiry into its Laws and Consequences* (1869), was perhaps the most influential.]

---

1 John Beddoe (1826-1909), physician, anthropologist, and author of the influential work *The Races of Britain: A Contribution to the Anthropology of Western Europe* (1885).

### Preference for narcotics

Many of them have been excessive in their abuse of narcotics, or of stimulants and intoxicants. Haller was in the habit of taking enormous doses of opium, and Rousseau was excessive in his use of coffee. Tasso was renowned as a drinker, and also the modern poets Kleist, Gérard de Nerval, Musset, Murger, Majláth, Praga, and Rovani, as well as the very original Chinese writer, Li-Tai-Pô, who was inspired by alcohol, and died of it. Lenau also, in his latter years, was an immoderate consumer of wine, coffee, and tobacco. Baudelaire abused opium, tobacco and wine. Cardan confessed himself an indefatigable drinker. Poe was a dipsomaniac; so was Hoffmann. (316)

### Genius and the love of difficulty

These energetic and terrible intellects are the true pioneers of science; they rush forward regardless of danger, facing with eagerness the greatest difficulties—perhaps because it is these which best satisfy their morbid energy. They seize the strangest connections, the newest and most salient points; and here I may mention that originality, carried to the point of absurdity, is the principal characteristic of insane poets and artists. (317)

### Egotism

All insane men of genius, moreover, are much preoccupied with their own *Ego*. They sometimes know and proclaim their own disease, and seem as though they wished, by confessing it, to get relief from its inexorable attacks. (319)

### Bipolarity

But the most special characteristic of this form of insanity appears to reduce itself to an extreme exaggeration of two alternating phases, viz., erethism[1] and atony,[2] inspiration and exhaustion, which we see physiologically manifested in nearly all great intellects, even the sanest—phases to which they, all alike, give a wrong interpretation, according as their pride is gratified or offended. (327)

---

1  Excitement of an organ or tissue to an unusual degree.
2  Relaxation or languor.

*Psychosis of genius*

Yet the temper of these men is so different from that of average people that it gives a special character to the different psychoses (melancholia, monomania, etc) from which they suffer, so as to constitute a special psychosis, which might be called the psychosis of genius. (329)

# Appendix B: Colonial Contexts: Accounts of the Indian "Mutiny," 1857-58

[In many senses, the Indian "Mutiny" of 1857-58, also referred to by historians as the "Uprising" and the "First War of Indian Independence," was the defining encounter between Britain and India in the nineteenth century. The sequence of uncoordinated uprisings began with the rebellion of sepoys, ostensibly protesting about the use of forbidden animal fat in greasing their cartridges, in Meerut on 10 May 1857 and continued more or less continuously until the spring of 1858. It was one of the most documented conflicts in history, prompting unprecedented levels of self-scrutiny about the colonial project at home and abroad. This appendix contains a representative sample of these reflections on the "Mutiny," with a specific reference to Agra in the summer of 1857 (the location of Small's narrative in *The Sign of Four*). The documents I have selected range from official history to personal memoir, and from church sermon to private correspondence, and provide an immediate and compelling insight, from a range of British perspectives, into the chaos and upheaval of that unforgettable Indian summer.]

1. **From Sir William Muir, *Agra in the Mutiny and the Family Life of W. & E.H. Muir in the Fort, 1857: A Sketch for their Children* (1896)**

[Previously a junior member of the Board of Revenue in Agra, William Muir (1819-1905) was head of the intelligence department inside Agra Fort during the 1857 "Mutiny." As well as maintaining the British contingent's official correspondence with Delhi, Cawnpore, and Calcutta, Muir also amassed intelligence notes, consisting of depositions from "informers, spies and messengers," from 5 July to 11 December 1857. Commissioned by Lord Canning to investigate claims of the mass rape of British women at the hands of the mutineers, Muir found no evidence to support this charge: "the almost universal opinion was that the attacks were purely murderous, with no attempt anywhere at dishonour" (57). Agra was effectively secured after the defeat of a contingent of mutineers from Indore on 11 October 1857. This account was compiled as a personal memoir of the events for Muir's children, in particular his daughter, Bessie, who was one of their five children to endure captivity in Agra Fort. Much of this material had earlier been used by Kaye in his official history of the Indian Mutiny, extracts of which are available here (Appendix B6). Muir was an accomplished (though decidedly partisan) scholar of Arabic, Persian and Islamic studies. He founded Muir College in Allahabad, as well as a small settlement, Muirabad, for Indian converts to Christianity.]

*Taking refuge in Agra Fort, 1 July 1857*

And thus the month of June [1857] wore away, till towards its close the Mutineer body from Nemuch and Nusseerabad,[1] instead of, as was expected, crossing over to Dehli, were found to be marching direct upon us. Then, as they approached, things began to look so threatening that the women and children were warned (it might well have been before) to take refuge in the Fort at once. The chamber allotted to us was a long, bare stone room on the lower floor on the east side of the Dewan-Khas, or Palace Square, with windows looking into the square. It served, when fitted with a few simple pieces of furniture, for all the purposes of the day, and for the numerous neighbouring friends who took their meals with us. At night the further end, when curtained off for our beds, was suitable enough for ourselves and the children. Beyond the actual necessaries of life and reasonable comfort, we could bring nothing to our room; but we managed to save and store away what things we chiefly wished to keep. It was Wednesday, the 1st of July, that your Mother, with our five children, left the dear old house at Hurree Purbut,[2] where we had lived so long, and occupied this room. (18-19)

*The view of Agra from inside the Fort, 5 July 1857*

Then followed the anxious watching from the ramparts of the Fort, the first intelligence of our force being obliged to retire, and the sad sight of the wounded brought in with the re-entering troops. Beyond one or two shots to keep the immediate neighbourhood clear, there was no firing from the Fort; all were now within the walls; beyond them, everything was in the insurgent's hands. From the ramparts we could see the bungalows far and near—the thatched roofs giving every

---

1 Garrison towns of the British Indian army, about 240 kilometres (150 miles) apart. Nimach is southwest of Delhi, between Gwalior and Udaipur; Nasirabad is in Rajasthan, close to the holy city of Ajmer. Sepoys in Nasirabad mutinied on 28 May 1857, and those in Nimach on 3 June 1857.
2 Named after a holy Hindu shrine in Srinagar, Kashmir, this is the district of Agra where Muir's family house was located.

facility to the incendiaries—in a vast blaze during all the night, and the savage Sowars[1] cantering round our flaming homes. Thank God for the Fort of Agra! What would it not have been for our dear ones on that dreadful night without it, but a place of awful peril! (21-22)

*Sikh prisoners replacing Indian guards*

Our position in Agra was in some degree complicated by having to guard our monster Jail by European troops, for the Jail nujeebs (armed guard) had gone off in a body towards the end of the month. If the 3rd Europeans should be required in the field, it would be necessary to draw off the men employed on this duty, and the only resource left was to make over the custody of the jail to the Sikh prisoners, who were to be released and armed for the purpose. (22-23)

*Racial and religious segregation*

The native Christians, several hundred in number, after some doubt as to whether there was room for them in the Fort, were all, thank God, allowed to enter, very much at French's[2] hands, for otherwise he would have stayed out with them, and have surely shared their fate. It was a noble act, which few would have attempted. Our servants were, with one exception (the mahometan[3] Bheestie[4]), faithful to us; but, of course, were not then allowed in the Fort, and in their absence the native Christians were of great use. But it required all the influence of your Mother and her friends to keep them from being harshly treated, for the unkindly feeling towards Natives had already begun to spread. (33)

*Punitive British reprisals, 8 July 1857*

We kept ourselves shut up in the Fort, though we had positively not a man to oppose us. On Wednesday (8th) a demonstration was made by marching a column through the city, and (I regret to say) by plundering the shop of a large Mahometan merchant in the military bazaar. (34)

---

1  Hindi, horsemen.
2  Captain Lucius John French, Commander of the 9th Lancers, killed outside Agra on 10 October 1857.
3  I.e., Muslim.
4  Hindi, water carrier.

And so we settled down, most of us with something however small to do, in an otherwise listless life, cut off from all the world, in a little world of our own, with no concern beyond the ramparts of our Fort. One remembers the strange feeling, as looking across the river, we felt that even the other bank was, one might say, not our own but foreign land. Yet, with a family and surroundings such as ours, there was still work (as Kaye says) for busy hands to do. It was during the worst hot months of the year that we were thus incarcerated, and it may well be imagined that, with none of the accessaries [*sic*] to moderate the heat, or avoid the floods of rain, it must often have been in our little quarters a wearisome time, especially for the little ones. In the lack of servants, we had not even bearers to pull the Punkahs,[1] and so with mosquitoes and other annoyances of the hot and rainy months, the surroundings were often stifling. (35)

## Intelligence work

It must have been a couple of days after the battle, that the Lieutenant-Governor placed me in charge of the Intelligence Department. This involved the very serious responsibility of keeping myself, by means of spies and informers, *au courant*[2] with the progress of the revolt in every direction. For this end, a body of confidential messengers had to be entertained and highly paid. Where the road was dangerous, as towards Cawnpore,[3] they had to carry little letters written on the thinnest paper, thrust sometimes into a quill or secreted in any part of the body. The risk was great, for they were often searched, and if any letter was found upon them, they were killed or even blown from guns; and it took from a week to a fortnight for a letter to get through to Cawnpore ... as the spies and informers came in at any hour of the day, and sometimes of the night, I used to take down their depositions from their lips at once; and the news, if important, was communicated to Mr. Colvin or other of the authorities. For some weeks that grand

---

1 Hindi, fan. The lack of Indian menial labour to pull the fans was a common British complaint during the particularly hot summer of 1857.

2 French, up to date, in the know.

3 Modern day Kanpur, a large garrison town on the right bank of the Ganges, halfway between Agra and Allahabad. One of the most infamous incidents of the Mutiny took place there on 15 July 1857, when some 200 unarmed British women and children were massacred and their bodies thrown into a well by the forces of the putative Maratha Peshwa, Dundu Panth (c. 1820-59), popularly known as the Nana Sahib.

old man, Choubey Gunshâm Doss, blind as he was, waited on me daily as my chief informant. Eventually he went away to watch matters at his Etah Tehseel, and there was killed, being surprised by the rebels. His brother, Jye Kishen Doss, was granted, both for his own services and in recognition of his brother's, the title of Rajah and C.S.I.[1] (37-38)

## 2. From Sir William Muir, *Agra Correspondence during the Mutiny* (Edinburgh: T & T Clark, 1898)

[This volume collects together letters written by William Muir to family members at home from Agra during the events of the "Mutiny." Muir notes the ransacking of treasuries by mutineers and opportunists alike in the first weeks of the uprising. Muir once again refutes allegations of the rape of British women at the hands of the mutineers.]

*On the causes of the "Mutiny": 18 May 1857*

It has long been known that our Native army—the Sepoys especially of the Regular line—was in an alienated state of mind, discontented and suspicious. This feeling, as you know, showed itself at Barrackpore[2] and elsewhere in the refusal to use cartridges believed by the Sepoys to be made with some objectionable stuff that would affect their caste. It has been doubted whether this was felt to be a real grievance, and not a mere blind to cover other objects or causes of discontent. I see no reason to doubt that it was felt to be a real grievance, and that the Government should have quietly and discreetly given in. The Sepoys are children. It was no use reasoning with them to show that there was nothing harmful of caste in the cartridges. (5-6)

*The "Mutiny" at Muttra: 2 June 1857*

Our last reverse has been at Muttra.[3] On Saturday last the 30th, a Company of the 44th Native Infantry from this went to relieve the Company of the 67th, which has been hitherto in charge of the Treasury there, and which it was intended should bring in here a portion of the accumulating treasure. Both Companies united (notwithstanding there had been a bad understanding between the two regiments

---

1  Companion of the Order of the Star of India.
2  Literally "the city of the barracks," a garrison town 24 kilometres (15 miles) from Kolkata, on the left bank of the River Hughli.
3  Now known as Mathura, a town on the Jamuna river some 50 kilometres (30 miles) northwest of Agra and 150 kilometres (90 miles) south of Delhi.

before), attacked their Officers, shooting one, plundered the treasure, and went off towards Dehli. (19)

*The recapture of Muttra, 5 June 1857*

Muttra has been reoccupied by Mark Thornhill and a few servants and Volunteers. All is quiet there; but advantage was taken by the bad characters about, on the mutiny of the Sepoys, to burn the bungalows and plunder all unprotected property. The ease with which a Magistrate and Collector with a handful of men recovers his authority *after the Sepoys have gone*, shows the nature of the rising as a military one, and the source of our difficulty. We have not even a handful of men to give out Magistrates and Collectors ordinarily to reinstate them. We have been hitherto so utterly and entirely dependent on our Sepoys. (29-30)

*On support for the relief of Delhi, 15 July 1857*

We have Dehli news up to the 8th. General Barnard died of cholera on the 7th or 8th. But the feeling of confidence was not impaired by the event. It was not mentioned who has taken command … the feeling in the City [Delhi] is that of disheartenment; they begin to think of aid from Dost Mohammed,[1] etc. Sikhs, Goorkhas[2] and Guides fight well. The 600 Sikhs in Dehli will no doubt turn in our favour when it comes to the assault. (36)

*On the role of Punjabi soldiers in suppressing the "Mutiny," 15 October 1857*

It was so odd to see Native soldiers about once again, and Native Sirdars.[3] The first feeling was to shrink from them as deadly enemies,

---

1  Amir Dost Muhammad Khan (1793-1863), ruler of Afghanistan from 1842 until his death in 1863. Previously their adversary, in January 1857 the British had signed a treaty of alliance with Dost Muhammad, mainly as a way of increasing the political pressure on their current adversary, Persia. The tactic worked, at least in the short term: in March 1857 the Anglo-Persian Treaty signed in Paris formalised a Persian withdrawal from Herat. Dost Muhammad's life is cleverly fictionalised in Philip Hensher's *The Mulberry Empire, or the two virtuous journeys of the Amir Dost Mohammed Khan* (London: Flamingo, 2002).
2  A hill people of Nepal, traditionally recruited by the British into the army. Then as now, the Gurkha regiments had a reputation for tenacity and ferocity.
3  Hindi, commanding officer (literally, "general").

but it was only a passing feeling called up by the memory of the enormities of our mutinous Bengal army. The open, smiling countenances of the Sikhs and Punjabies[1] at once dispelled all such ideas. They are noble fellows. (42)

### 3. From James P. Grant, *The Christian Soldier: Memorials of Major-General Sir Henry Havelock* (London: J.A. Berger, 1858)

[Published as a cheap potted biography of Havelock and sold for 1s, Grant's *Memorials* had an explicitly Evangelical and populist appeal; in the preface, he dedicated his work to "the British public, to shew that the faith of a Christian is not incompatible with the duty of a soldier." Grant uses extensive quotations from other published sources on Havelock and the "Mutiny."]

*On the Christianisation of the army*

Happy is it for England that her army and navy is permeated with the leavening influence of the Gospel of Jesus Christ; so that even the world is constrained to bear witness that it is *not* necessary in order to make a "good Christian" that you should spoil "a good soldier." ... One great object which Sir Henry Havelock ever had in view was the Christianisation of the army; for this he laboured with untiring and self-devoted perseverance. (xi)

*On the suppression of the "Mutiny" as a just war*

However we may and do deplore war in the abstract, we must at least allow that in this case it was the stern necessity of justice—a case in which mercy could have no part. That moveable column might well be called the *avenging column*—like Gideon[2] of old, he had to fight against a people accursed of God for their horrid crimes. His mission was one peculiarly fitted for his puritan-like zeal; and if ever war was a pursuit lawful to the Christian, it was that war in which he was now engaged. (31)

---

1  Inhabitants of the Punjab, a linguistically homogenous region, whose population was divided between the Hindu, Muslim and Sikh religions. Muir's use of "Punjabies" in addition to the term Sikh here clearly indicates new recruits from the Punjab who were *not* Sikhs.
2  Biblical King and warrior in the Book of Judges, Gideon was a destroyer of idols.

*On British retribution for the massacre of civilians at Cawnpore*

When a rebel is caught, he is immediately tried, and unless he can prove a defence, he is sentenced to be hanged at once; but the chief rebels or ringleaders I [General Neill, the commander at Cawnpore] make them first clean up a certain portion of the pool of blood, still two inches deep, in the shed where the fearful murder and mutilation of women and children took place. To touch blood is most abhorrent to the high-caste natives; they think by doing so they doom their souls to perdition. Let them think so. My object is to inflict a fearful punishment for a revolting, cowardly, barbarous deed, and to strike terror into these rebels. The first I caught was a subadar, or native officer, a high-caste Brahmin, who tried to resist my order to clean up the very blood he had helped to shed; but I made the provost-marshal do his duty, and a few lashes soon made the miscreant accomplish his task. When done, he was taken out and immediately hanged, and after death buried in a ditch by the road-side. No one who has witnessed the scenes of murder, mutilation and massacre, can ever listen to the word "mercy" as applied to these fiends. (52-53)

*On India as a testing ground for the British character*

His victories were not the natural result of the opposition of European troops to Hindoos,[1] but were achieved by his own skill, courage, coolness, and indomitable energy ... more than any other man engaged in India he fixed the attention of his countrymen at home. It is probable in conjectures like this that a man's previous character most stands him in good stead. When a man has passed through life with blameless character, inoffensive manners, and evincing always a strong sense of duty, he has laid up for himself a treasure in the esteem of others of which, when the time comes, the interest is returned to him in full. It is possible that if India had never been in revolt, Colonel Havelock might have gone to the grave with only the reputation of a meritorious officer and a good man; but when once he had an opportunity of exhibiting his greater qualities, all the goodwill of those who knew him added to his renown, and their descriptions of what he was went to form that ideal which his countrymen conceived of his character. (101)

---

1  Used generically here to mean Indian, rather than specifically the Hindu population.

## 4. From Rev. Frederick S. Williams, *General Havelock and Christian Soldiership* (London: Judd & Glass, 1858)

[This is the printed text of a sermon delivered by the Reverend Frederick S. Williams of Birkenhead on 17 January 1858. It is a representative example of the posthumous elevation of the standing of Havelock, who, more than any other military commander, became a vehicle for the Evangelical movement. Havelock himself was a committed Baptist and the son-in-law of the famous Serampore Baptist missionary printer, Joshua Marshman (1768-1837).]

*On the news of the "Mutiny" at home*

Some eight months have passed away since there flashed forth upon a hundred dials the telegraphic[1] announcement, that the fairest and costliest jewel was falling from the crown of England. India, with its strange history, once colonized by a company of traders, now the noblest appendage ever held by state or potentate, and the richest dominion on the globe; India, so vast in its resources, so mighty in its influence, the possession of which had brought to this country so much power and prestige, the loss of which would have reduced her at one fell swoop in the scale of nations, was in arms against the crown and rule of England. Taken by the sword, she had taken the sword. The tidings awoke universal anxiety. Their novelty startled; the atrocities they recounted enraged. Some at once desponded, and declared that the sceptre of the East was dross; others looked hopefully for the immediate suppression of the revolt, and predicted that the flames of rebellion could soon be trodden out or quenched in blood; all waited with solicitude the arrival of fresh intelligence, and when it came, read or heard it with bated breath. But if the fears of alarmists were not fulfilled, neither were the hopes of the sanguine. The mutiny had spread, anarchy was rife, regiment after regiment had murdered its officers in cold blood, and then gathered to the standard of revolt; twenty, thirty, forty thousand men, fed with our food, armed with our arms, disciplined with our discipline, were in fierce array; Delhi, the ancient capital of the Moguls, was the head-quarter of the mutineers; Calcutta was in danger. Darkly gathered the clouds over the land, and as the

---

1  First demonstrated by Samuel Morse in 1844, the telegraph brought a new immediacy to the events in India to the British public. The Electric Telegraph Act ("an Act for regulating the establishment and management of Electric Telegraphs in India") was passed on 23 December 1854, and by the summer of 1857, the new technology had become an indispensable source of transmitting information.

flames of rebellion found dry and rotten fuel on every hand, they spread near and far, and lighted up the heavens with their lurid, ghastly hue. (3)

## The British response and the construction of heroism

England gathered up her strength. Thirty vessels shook their canvass to the breeze, and borne upon the wings of the wind the van of the avenging host left her shores, and soon more than forty thousand men were hastening to the seat of war. But while the brow of England darkened at a peril so imminent, while her blood grew hot at the atrocities perpetrated against her sons and daughters, which made her doubt whether they had not been enacted by fiends who, in some fell hour, had flung back the portals of perdition, and broken loose from hell, forth with her strong right arm flashed the falchion[1] from its sheath, and held high the sword of justice glittering in the light of heaven, not to be held in vain. Meanwhile tidings reached her of the deeds of heroism, of the martyr-like devotion of her children in that fearful strife. Every Englishman in India seemed to be a hero. But the highest officers to whom she confided the guidance of her forces fell one by one in that unequal struggle. Anson, Barnard, Henry Lawrence, "the fiery Neill," "the polite Nicholson,"[2] were overborne by death or disease; and then was heard again and yet again the name and deeds of whom we have to make especial reference to-night. (3-4)

## Havelock's character and leadership

The "slight, spare figure" of the general—for he is but five feet five in height, and looks as if a week's exposure would break him down—the grave look, the emaciated face, the almost unobservant apathetic expression, excites little to awaken admiration or reverence; but the steel is burnished, and the blade is keen, that rests within that seemingly frail scabbard. The boy whose gravity awoke the playful sallies of

---

1  A broadsword with the edge on the convex side.

2  General George Anson (1797-1857), commander-in-chief of the British Army in India, died of cholera in Karnal on 27 May 1857; Sir Henry William Barnard (1798-1857), died of cholera outside Delhi on 5 July 1857; Henry Montgomery Lawrence (1806-57), killed during the attempt to relieve the siege of the Lucknow Residency on 4 July 1857; James George Smith Neill (1810-57), killed also at Lucknow, 26 September 1857; John Nicholson (1821-57), veteran of the First Anglo-Afghan War, killed at the storming of Delhi, 23 September 1857. All the officers listed here were known for their publicly espoused and proselytising Evangelical beliefs.

his schoolfellows at the Charterhouse,[1] as passed into a quiet, calm, undemonstrative man; yet one of deep conscientiousness, profound feeling, and iron tenacity of purpose. But the moment for action comes. An aide-de-camp arrives at a gallop with tidings that the enemy is posted in force in front—that in deep, dark masses he crests the ridge and occupies the village, making each house a fortress garrisoned; and that the roads are commanded with heavy guns. Now you shall see the man—the mind. The sword leaps from its scabbard. The eagle eye scans each position. Instinctively the General assumes the air and bearing of command; instinctively, all obey. He has never lost a battle yet, and the men are sure he never will. He grows quite chatty and agreeable as the shot whistles, for the guns of the enemy have found the range and are well plied. The plan of attack is swiftly conceived. Skirmishes are thrown out in front, and aides-de-camp bear their instructions to the officers. Now for a moment that little army pauses, like some swift eagle poised upon the wing, and then with lightning swiftness it swoops upon its quarry, tosses resistance to the winds, and nestles sternly, with blood-stained bosom, upon the heights from which it has hurled, or on which slain, the recreant foe.

The work accomplished, the general relapses to his former quietude; the flame, before so beacon-like, must not be frittered away in needless excitement; and sinking and slumbering amid the embers is gathered up for future, and perhaps early use. (5)

*The idea of the "Christian Soldier"*

Henry Havelock enlisted in two armies: a carnal and a spiritual. He served under two sovereigns; the Queen of England and the King of kings. He fought in two campaigns: against the foes of England, and against the world, the flesh, and the devil.—It has been remarked by those who have visited the church-yard and the field of Waterloo,[2] that there is hardly one epitaph which breathes over the departed any hope of a resurrection—hopeful as epitaphs generally are; but the graves of

---

1 Founded in London in 1671, a famous English public school that offered scholarships to relatively poor but academically promising boys. Based in Godalming, Surrey since 1872, alumni include the novelist William Makepeace Thackeray (1811-63) and the founder of the scout movement, Robert Baden-Powell (1857-1941).

2 Decisive British victory, on 18 June 1815, in the Napoleonic Wars.

the Crimea[1] tell a different tale ... like most of our Indian statesmen and soldiers, the Lawrences, Edwardes,[2] Nicholsons, Montgomerys, and many others, Henry Havelock was a Christian of the old stamp—a strong God-fearing Puritan man, who thought often in Scriptural phrases, and deemed it no shame to teach his soldiers to pray. (6)

## 5. From Mrs. R. M. Coopland, *A Lady's Escape from Gwalior and Life in Agra Fort during the Mutinies of 1857* (London: Smith, Elder & Co., 1859)

[She was the widow of the Reverend George William Coopland, a chaplain to the East India Company resident at Gwalior, who was murdered during the "Mutiny." Forced to flee the city during the uprising, she joined the swelling ranks of the British sheltering inside Agra Fort. Mrs. Coopland's account was published by Smith, Elder & Co. in March 1859 in a single octavo volume priced at 10s 6d, pitched at both lending libraries and individual purchasers; she provides one of the most immediate (if not the most accurate) accounts of the events of 1857-58. Coopland also includes her (and her husband's) letters home during this period as documented testimony to historical events, and she recounts an interview with the defeated and imprisoned last Mughal Emperor, Bahadur Shah Zafar (1775-1862), before his exile to Burma. *A Lady's Escape* demonstrates the tremendous appetite for personal memoirs about the "Mutiny" in post-1857 Britain, as well as the hardened, largely Evangelical intolerance, that many such memoirs fed off, and indeed fostered.]

*The cause of the "Mutiny"*

It seems surpassingly strange that so little notice was taken of the impending danger by those whose duty it was to care for the safety of a mighty empire. We had, at the beginning of the year 1857, three regiments *less* than before the annexation of Oude.[3] There were no Euro-

---

1 The Crimean War (1853-56), which pitted an alliance of Britain, France, and Ottoman Turkey against Tsarist Russia. Britain had entered the war in a joint expeditionary force with the French in May 1854. Defeat at Balaclava had shaken British resolve, but the conflict was eventually settled, partly inconclusively, on 30 March 1856.

2 Herbert Benjamin Montgomery Edwardes (1819-68), proponent of the Anglo-Afghan treaty with Dost Muhammad and a committed proselytising Evangelical, held Peshawar during the "Mutiny."

3 Also spelled Oudh or Awadh, a self-governing Indian state with its capital at Lucknow. The annexation of Oudh in February 1856 was another of the injustices that fuelled the uprising in 1857.

pean regiments at many of the largest stations: Allahabad, Cawnpore, Benares, and Delhi, were all left to the protection of disaffected regiments. The Government at Calcutta, in serene complacency, was coolly issuing orders for the disbanding of regiments: as though that could in any way stop the evil.

We now heard of the hanging of Mungul Pandy[1] and of incendiarisms at Umballa.[2] Many reasons were assigned for these disturbances: first, the trumpery one of the greased cartridges; and, secondly, the annexation of Oude. But neither of these were the *real* reason. (79)

*Rev. Coopland's explanation for the "Mutiny"*

This is God's punishment upon all the weak tampering with idolatry and flattering vile superstitions. The Sepoys have been allowed to have their own way as to this and that thing which they pretended was part of their religion, and so have been spoiled and allowed to see that we were frightened of them. And now no one can tell what will be the end of it. There is no great general to put things right by a bold stroke. We shall all be cut up piecemeal. (85)

*The insubordination of servants during the "Mutiny"*

I was much struck with the conduct of our servants—they grew so impertinent. My ayah[3] evidently looked on all my property as her share of the plunder. When I opened my dressing-case, she would ask me questions about the ornaments, and inquire if the tops of the scent-bottles were real silver; and she always watched where I put my things. One evening, on returning from our drive, we heard a tremendous quarrelling between the Sepoys of our guard and the ayah and khitmutghar.[4] They were evidently disputing about the spoil; and it afterwards turned out that the Sepoys got quite masters, and would not let the servants share any of the plunder, but kept them prisoners,

---

1  Mangal Pande, also spelled Pandey (1827-57), first leader of the mutinous sepoys in Barrackpore (near Calcutta), was executed on 8 April 1857. A memorial to him as the leader of India's "First War of Independence" is located in Calcutta's BBD Bagh, opposite the Writer's Building. For two interesting recent interpretations of Mangal Pande, see Amaresh Misra, *Mangal Pandey: the true story of an Indian revolutionary* (New Delhi: Rupa & Co., 2005) and Rudrangshu Mukherjee, *Mangal Pandey: Brave Martyr or Accidental Hero?* (New Delhi: Penguin India, 2005).
2  Ambala, a large garrison town in the Punjab.
3  Hindi, female domestic (usually a nanny or nurse).
4  Persian and Hindi, butler.

and starved and ill-treated them. They had much better have remained faithful to us, and have helped us to escape; instead of which, at the first shot, they vanished, and began to plunder what they could. My husband overheard the punkah coolies outside talking about us, and saying that these Feringhis[1] would soon have a different home, and *they* would be masters ... I could not help fancying they might have made us punkah and fan *them*, so completely were we in their power. (109-10)

*Description of the mutineers in Gwalior*

Oh! how could the bright sun and clear blue sky look on such a scene of cruelty! It seemed as if God had forgotten us, and that hell reigned on earth. No words can describe the hellish look of these human fiends, or picture their horrid appearance: they had rifled all the stores, and drank brandy and beer to excess, besides being intoxicated with bhang.[2] They were all armed, and dressed in their fatigue uniform. I noticed the number on them; it was the 4th—that dreaded regiment. Some were evidently prisoners who had been let out from the gaol the night before; and they were, if possible, more furious than the rest ... the road was crowded with Sepoys laden with plunder, some of which I recognised as my own. (127)

*Treasure buried during the "Mutiny"*

It is a common practice with natives, when there is an alarm, to bury their property; and in this way most of the Delhi loot was found. When a native suspects treasure has been buried, he searches for the place where he thinks it has been concealed, and throws water on it; if it sinks rapidly in, he knows the earth has been recently stirred, and then begins digging; and if he is lucky, he may light on some earthen vessel filled with gold mohurs[3] or rupees. I was told that Major Hodson, of Delhi renown, once, on a foraging expedition, came to a wall which his keen eye perceived to be the depository of treasure. He instantly

---

1  Hindi, foreigner. Originally specifically used for the French, but by this time, a generically derogatory term for all Europeans.

2  The narcotic, Indian hemp (*Cannabis sativa*). Often consumed as a drink, sometimes blended with milk, yogurt, or opium suspension, bhang was sometimes used in Hindu religious rituals.

3  A Standard gold coin unit in British India, the mohur was in official circulation from 1831 to 1895, although it continued to be used in self-governing territories in India until the 1930s. The gold mohur was usually valued at 15 rupees, or just over £1.

went in search of assistance; but on his return, to his great disappointment, he found nothing but empty vessels: some one having been too quick for him had carried off the golden store. (146)

*Description of Agra Fort*

We descended a steep hill, and the fort then loomed upon us in all its massive strength, with its walls and battlements of dark red stone, and its formidable looking entrance guarded by some of the 3rd Europeans. We had not time then to estimate its merits and defects; and though its massive walls and loopholes, from which frowned the cannon, and the gateway with its drawbridge spanning the wide and deep moat surrounding the fort, were assuring, still we could not help shuddering at the possibility of its being besieged. Indeed, had the enemy then attacked it, our small force would have been quite insufficient to defend its immense extent of walls and ramparts in such an exposed position; and there were afterwards found to be numerous underground passages, leading from the city immediately into the fort. We can now look back with thankfulness that we were not exposed to a siege. (159)

*The inhabitants of Agra Fort*

The half-castes, or "Kala-Feringhis," as the natives called them, who are uncharitably said to have the vices of both different races, and the virtues of neither, were in immense swarms, and had to accommodate themselves anywhere. A large number of them lived in our "square," just beneath our balcony: the rest lived in holes, tyrconnels, or on the tops of the buildings all over the fort. Poor creatures! they must have had a miserable time of it; for their habitations were very wretched. The census of all the persons in the fort, which was taken on the 26th of July, amounted to no less than 5845; of which 1989 were Europeans, consisting of 1065 men and 924 women and children: the whole of the rest being natives and half-castes. (171-72)

*British officers looting Agra*

The officers sometimes made parties to go into the city and loot; but so great was the devastation, that they never brought us back anything, except a few cups and saucers and a coffee-pot. They told us it was the most wretched and forlorn sight they had ever seen; nothing but the charred walls of houses, with furniture, books, and pictures, utterly destroyed, lying about the streets. (186)

*Paranoia within Agra Fort*

The manners of the servants were most insolent and contemptuous; they often said that our "rajh"[1] was over, and considered us doomed; fully expecting that when their brethren had defeated us in Delhi, which they never doubted would be the case, they would march to Agra and cut us all to pieces with little trouble ... they would also often lie down in our rooms, and when we spoke to them, did not get up. The "budmashes"[2] used to sing scurrilous songs under the walls, and draw pictures on them of the "Feringhis" being blown up, with their legs and heads flying into the air; they also stuck up placards, saying on such a day we should all be massacred or poisoned. One baker was really hanged for planning a scheme for poisoning all the bread; and it was feared they might poison the wells ... rumours of a depressing nature from Delhi, news of fresh mutinies, and massacres of men, women, and children, daily poured in; and the tidings from England were that, instead of sending out troops "overland" instantly, the parliament and ministers were disputing and squabbling among themselves over the causes of the mutiny, and weighing the comparative merits of greased cartridges and cow's fat, forgetting the fearful loss of life going on in the meantime. (188-89)

*On the newly recruited Sikh soldiers in the British Army at Agra*

Colonel Greathed left us a guard of 200 Seiks, wild, savage-looking men, and so ragged and dirty, they reminded me of gipsies. They had most curiously shaped swords, and wore queer sort of headgear. They said it was very hard that they were not allowed to "loot" Agra, as it was such a rich city ... we heard of a Seik finding some jewels of great value; but I don't know whether the report was true. Many said the Seiks found a heap of plunder; but not so much as was expected, the Sepoys having taken a great part of the "loot" with them when they fled. (231-32)

*British retribution in Agra after the suppression of the "Mutiny"*

Many prisoners were hanged after the battle, and as it was discovered they did not care for hanging, four were tried and sentenced to be blown from guns; accordingly one day we were startled by hearing a gun go off, with an indescribably horrid muffled sound. We all rushed

---

1 Hindi, rule or sovereignty.
2 Hindi, scoundrels.

out of our "dens" to know what was the matter; and heard that some Sepoys were being blown from the guns. All our servants hurried away to see the sight; and then was heard, at short intervals, three other guns go off. The sound was horrible, knowing as we did that a fellow creature (whatever he may have done) was being blown into fragments and his soul launched into eternity at each report of the cannon; and we felt quite ill for the rest of the day. An officer told us it was a most sickening sight. The four guns, taken out of the fort, were placed near the river. One gun was overcharged, and the poor wretch was literally blown into atoms, the lookers on being covered with blood and fragments of flesh: the head of one poor wretch fell upon a bystander and hurt him. It was a long process, fastening them to the guns; and an officer having said to a sepoy, as the latter was being tied on, "It is your turn, now," the sepoy replied calmly, "In one moment I shall be happy in Paradise." (233)

*Celebrating the British recapture of Agra*

In the second week in December, the wounded soldiers, who had now recovered, gave a grand fête at the Taj, in honour of the ladies who had attended them. Regular cards of invitation and programmes were printed; the tiffin[1] was to be spread in one of the mosques, and nearly every one went to it ... the road was crowded with soldiers and carriages, and the river with boats, all on the way to the Taj. It was a very gay scene. In one of the mosques of the Taj, all the ladies, children, officers, and soldiers were gathered; and here and there might be seen a native, looking green with rage at their sacred building being thus desecrated. The mosque was beautifully decorated with flowers, and a table was spread with all the dainties that could be procured. Almost everyone looked happy and cheerful, and the ladies went from one soldier to another saying kind words, and congratulating them on their recovery.

We staid here and had some tiffin; and I remember one man of the 9th Lancers offering me some milk punch,[2] that being the only beverage they could procure, as the wine in the fort was drunk. (245)

*On the loot in Delhi palace*

Captain Garstone again came, and drove me to the Palace, which is surrounded by lofty walls of red sandstone between fifty and sixty feet

---

1  An Anglo-Indian word, usually referring to a mid-morning meal.
2  Despite its innocuous sounding name, traditionally an Anglo-Indian beverage made with two parts milk to one part brandy, mixed together with lemon juice, orange juice and sugar.

high, at the opposite end of the Chandney Chowk.[1] Before the barbican were patrolling a man of the 60th Rifles, that gallant regiment which had done such good service in the siege, and a diminutive Ghoorka, with a grotesquely solemn face, looking too small to shoulder his musket. We drove into a small courtyard, and then through a magnificent gateway into a long and lofty arched corridor. Here, in small recesses on each side, lived little Ghoorkas, who were cleaning their arms or smoking: their downcast looks harmonised with the gloomy solitude of this once luxurious place. We turned out of this ante-like corridor down a narrow road, leading to the small mosque, which was surrounded by tumble-down, squalid-looking buildings.

I received a kind welcome from Mrs. Garstone, who was much pleased with my baby; and they gave me one of the three compartments, into which the mosque was divided by purdahs.[2] These purdahs looked like arras, and made me feel as if we had gone back to the old tapestry days; the effect was carried out by a large fire blazing on a brick hearth, which Captain Garstone had contrived out of one of the sacred recesses, some antique chairs, and a table and shelves. On these were arranged some valuable "loot" which Captain Garstone had found in the king's palace; including a beautifully bound and illuminated Koran on vellum; a curiously carved sandal-wood cane, in which was concealed a dagger; a strange picture of some old Mogul, very like the present king; a splendid casket of ebony and mother of pearl, full of secret drawers (which Captain Garstone was very much disappointed to find contained no jewels); a solid silver enamelled flask for scent, and a variety of charms, bangles, signet-rings &c. (257-58)

*Mrs. Coopland's "final solution" for Delhi*

We soon took a last view of the "City of Horrors." I could not but think it was a disgrace to England that this city, instead of being rased [sic] to the ground, should be allowed to stand, with its blood-stained walls and streets,—an everlasting memorial of the galling insult offered to England's honour. Many would forget this insult; but it cannot, and ought not to be forgotten. Yet the natives are actually allowed to ransom back their city, street by street; whereas, if it were destroyed, being their most sacred city, and one that reminds them of their fallen grandeur, it would do more to manifest our abhorrence of their crimes, and our indignation against them, than the hanging of hundreds.

---

1  Literally "moon street," one of the main thoroughfares in Old Delhi.
2  Hindi, screen.

Delhi ought to be rased [*sic*] to the ground, and on its ruins a church or monument should be erected, inscribed with a list of all the victims of the mutinies,—if it be possible to gather the names of ALL those who were massacred,—and the funds for its erection should be raised by a fine levied on every native implicated in the mutinies, but not openly accused of murder.

Not only our victories of 1857 must be remembered, but the cruel massacres of English men and women which preceded them. Such atrocities ought never to be buried in oblivion. (278-79)

*On the deportation of mutineers to the Andamans*

We had to wait some time at the landing-place [Karachi port]; as a vessel which was anchored some way out was waiting for its cargo of Pandies,[1] ready to take them to the Andaman Islands. The sepoys, chained together in couples, and manacled, were coming down the steps into the boats. They looked a wretched, miserable, dirty set, and the clanking of their chains had a dismal sound. The captain of their vessel, a Yankee, said he would "break them in." They were to clean out their "dens" or "hutches" on board, and eat bacon or anything, regardless of caste.[2] Some people said that few would reach their destination, as they suffer so much from sea-sickness and have such a devout horror of the "black water," and have been known to mutiny rather go by sea from Calcutta to Bombay, or elsewhere.

The Andaman Islands, a group of four islands inhabited by savages, in the Bay of Bengal, are so unhealthy, that though the English tried to form a settlement there in 1791, they were obliged to abandon them in five years. It was said that each <u>sepoy</u> was to be allowed a knife, to defend himself against the savages, and some food; and if he behaved well, his wife was to be sent to him in two years. (303-04)

## 6. From Sir J.W. Kaye and G.B. Malleson, *The History of the Indian Mutiny of 1857-8* (London: W.H. Allen, 1888-89), 6 vols

[Sir John William Kaye and Colonel George Bruce Malleson's comprehensive six-volume history of the Indian Mutiny was considered to be authoritative at the time, and was the culmination of both authors' accounts of the events of 1857-58. The first two volumes were written by Kaye as *A History of the Sepoy War in India, 1857-8*, the remainder

---

1  A generic term for mutinous sepoys, after their first leader, Mangal Pande.
2  Forbidden to both Muslims and caste Hindus on the grounds of religious dietary law.

by Malleson (Kaye died in 1876). The extracts reproduced here are from the Cabinet Edition (6 vols.) published by W.H. Allen & Co., London, from 1 October 1888 to 1 October 1889. This would be a reference work that many of the putative readers of *The Sign of Four* would have known. Kaye and Malleson's style is readable, and while partisan, the tone is rarely triumphalist; they are, for example, explicit in their condemnation of the execution of the sons of the Mughal Emperor Bahadur Shah Zafar (by Hodson) after their unconditional surrender to the British on 23 September 1857 (Vol. 4, 52-57). All of Kaye's first volume exhumes the prehistory of relations between India and the British leading up to the "Mutiny," examining incidents such as the Vellore mutiny (1806) and the increasingly Evangelical tendency of British officers through the course of the nineteenth century.]

*The construction of Englishness through the events of 1857*

The story of the Indian Rebellion of 1857 is, perhaps, the most signal illustration of our great national character ever yet recorded in the annals of our country. It was the vehement self-assertion of the Englishman that produced this conflagration; it was the same vehement self-assertion that enabled him, by God's blessing, to trample it out. It was a noble egotism, mighty alike in doing and in suffering, and it showed itself grandly capable of steadfastly confronting the dangers which it had brought down upon itself. If I have any predominant theory it is this: Because we were too English the great crisis arose; but it was only because we were English that, when it arose, it did not utterly overwhelm us. (Vol. 1, xi)

*Accounts of the disorder in Agra, 6 July 1857*

Before the survivors entered [Agra Fort], the blaze, advancing from house to house in the cantonments and civil station, had told the non-combatants and ladies within the fort how the battle had been appreciated by the natives. Hordes of villagers who had watched the contest from afar had at once dispersed to burn and to plunder. The previously released prisoners, and their comrades, now set at large, joined in the sport. All night the sky was illuminated with the flames of burning houses, and a murmur like the distant sea told what passions were at work. It was a magnificent though sad spectacle for the dispirited occupants of the fort.

During the two days following disorder was rampant in and outside the fort. The city, the cantonments, the civil lines were ruthlessly plundered. Of all the official records those only of the revenue department were saved. (Vol. 3, 185-86)

*British demonstration of power, 8 July 1857*

The following morning he [Mr. Drummond, the British magistrate of Agra] issued from the fort, escorted by a company of Europeans and some guns, made a circuit of the principal streets and of the station, and proclaimed the restoration of order and British rule ... the restoration of order in the fort followed Mr. Drummond's action in the town. The natives of the lower class, prompt to appreciate decision, returned as if by magic to their duties. Prior to Mr. Drummond's triumphant tour through the city, there had been a great dearth of servants in the fort; but the day following small shopkeepers flocked in with provisions; domestics of every grade were eager to renew or proffer service ... the natives had seen the utmost the rebel troops could accomplish; and their faith in British ascendancy revived. (Vol.3, 186-87)

*British preparations for the defence of Agra Fort, September 1857*

To combat the facts and rumours surging about him, Mr. Reade, in conjunction with Lieutenant-Colonel Cotton, commanding the garrison, issued orders, on the 19th September, to set to work at once to level some obstacles which interfered with the free play of the guns mounted on the fort, and to mine some of the more prominent buildings, including the great Mosque, which were in dangerous proximity to the walls. (Vol. 4, 67)

# Appendix C: Colonial Contexts: The First and Second Anglo-Afghan Wars

[British Imperial policy in India was shaped in large measure by its engagement, through both violence and diplomacy, with its north-western neighbour and buffer state, Afghanistan, a policy of both confrontation and containment that became known as "The Great Game." The British fought three separate wars with Afghanistan over a period of 80 years (1839-42, 1878-80, and 1920) before finally conceding the independence of Afghanistan as a neutral state through the signing of the Anglo-Afghan Treaty on 22 November 1921.

The First Anglo-Afghan war was precipitated by the overthrow of Shah Shuja-ul-Mulk Saddozai (1785-1842) by Dost Muhammad Khan (1793-1863), who crowned himself Amir of Afghanistan in Kabul in 1836. In October 1838, Lord Auckland, the British Governor-General of India, issued the Simla declaration, which demanded that Shah Shuja (who was amenable to British interests) be reinstated to the throne, by force if necessary. In November 1838, an expeditionary force, the Army of the Indus, set off to enforce the Simla declaration; in April 1839, the British captured Kandahar, and before the end of the summer, the British had entered Kabul and reinstated Shah Shuja. In 1840, Dost Muhammad Khan surrendered to the British and was sent to exile in India, but in November 1841, a widespread insurrection in his favour broke out in Kabul. The British political resident in Kabul, Sir Alexander Burnes, was assassinated, Shah Shuja deposed and imprisoned, and the British envoy and minister, Sir William Macnaghten, killed at a conference with Afghan chiefs in December 1841. The British in Kabul were besieged, and some 4,500 British and Indian troops, together with 12,000 camp followers, fled the city for Jalalabad and the border; only one European survivor reached Jalalabad alive. In February 1842, Shah Shuja was murdered, and in September a second British expeditionary force successfully relieved Jalalabad and then Kabul. In 1843, Dost Muhammad Khan returned to Kabul in triumph, and the humiliating setback forced the British to accept his de facto (if not de jure) authority. By the time of the Indian "Mutiny" in 1857, skilful diplomacy had managed to enlist Dost Muhammad Khan on the British side in that conflict.

Things did not remain stable after Dost Muhammad Khan's death in 1863; his son and successor, Amir Sher Ali Khan, proved to be a divisive figure who plunged the country into another civil war (two of his sons rebelled against him and went into exile). Fearing Sher Ali's increasingly pro-Russian sympathies (cemented through the establishment of a Russian mission in Kabul), in 1878 the Viceroy of India, Lord Lytton, issued an ultimatum demanding that he accept a British mission to Kabul. In August 1878, the British mission was turned back at the border, thereby precipitating the Second Anglo-Afghan War. In November 1878, three British columns entered Afghanistan, bound for Kabul, Jalalabad, and Kandahar, and forced Sher Ali to flee to Mazar-i-Sharif, where he died in February 1879. Just weeks after the British had established a residency in Kabul in the summer of 1879, mutinous Afghan soldiers rose up and murdered the British resident and all his staff; by December 1879, the "mutiny" had become a widespread tribal rebelling across the country. In an attempt to legitimise their military occupation, the British proclaimed Abdur Rahman Khan Amir of Kabul in July 1880, but this did not pacify the country, and later that month (27 July) Muhammad Ayub Khan decisively defeated the British at the Battle of Maiwand. In turn, the British under General Roberts inflicted a defeat upon Muhammad Ayub's forces in September, but unable to hold the territory they had captured, the British were forced to withdraw later that year, and in April 1881 they also withdrew from their last remaining stronghold, Kandahar. The Second Afghan War ended inconclusively, but the dispatch of a new British delegation to Kabul in 1882 suggested that both sides were prepared to engage, for the moment, in diplomatic dialogue. At the time Doyle wrote *The Sign of Four*, the Afghan Boundary Commission had just finished delineating the border between the two countries, known as the Ridgeway Line. This British designated border has continued to be an issue of contention between Afghanistan and Pakistan.

The extracts below are largely personal accounts of the First Anglo-Afghan War, and newspaper reports of the aftermath of the British involvement; the report from *The Times* on the murder of Lord Mayo by an Afridi tribesman while visiting the penal settlement on the Andamans demonstrates the awareness of contemporary British readers of the potential for retribution harboured by Afghans who resented British interference in their affairs. Lady Florentia Sale's account, one of the most compelling accounts of conflict in the nineteenth century written by a woman, is an exemplary narrative of the disastrous First Anglo-Afghan War. It provides a literary model (a narrative of captivity, escape, and survival) that would emerge and proliferate in British women's accounts of the Indian "Mutiny" fifteen years later. Disquiet over relations with Persia, the territorial ambitions of Russia, and potential conflict with Afghanistan structured the British colonial (and especially, British Indian) imagination through the nineteenth century. Watson's injury, picked up at the Battle of Maiwand in the Second Anglo-Afghan War, is more than merely a trivial anecdote, for it suggests once again both Doyle's and his readers' understanding of the commitment of men and materiel to an unresolved Imperial policy. The Afghan wars continued to haunt the British Imperial consciousness, and like the collective memory of the Indian "Mutiny," Doyle's readers would have been only too aware of its implications for the continuation of British rule in India.]

## 1. From Sir Henry Havelock, *Narrative of the War in Affghanistan, 1838-9* (London: Henry Colburn, 1840)

[Before he became a general and the posthumously knighted subject of hagiographies for his actions during the "Mutiny" of 1857-58, Henry Havelock had established his reputation as one of the few British army officers to emerge from the disastrous First Anglo-Afghan War unscathed. At the time a captain in the 13th light infantry regiment, and serving as aide-de-camp to Sir Willoughby Cotton (1783-1860), commander of the Bengal forces in Afghanistan, Havelock's account deals only with the first half of the conflict, i.e., the advance and *not* the disastrous retreat of 1841-42. Combining military detail with a careful eye for picturesque travelogue, Havelock's account provides a remarkable account of Imperial overconfidence; despite his literary ambitions, the book was not a sales success. Havelock was officially commemorated by an 1861 bronze statue by William Behnes which stands in Trafalgar Square.]

*On the psychological importance of British demonstrations of strength*

One hour of such success outweighs the results of months of intrigue and negotiation; and that the clash of steel for a few short moments will ever gain for the British, in the diffusion through Asia of an opinion of their strength, a greater advantage than all the gold in their coffers can purchase. (I, 335-36)

*Sikhs either to be co-opted or crushed to facilitate the Afghan policy*

Henceforth there can be no medium, therefore, in the character of our relations with the Sikhs; they must either be established on a footing of the closest intimacy, or change at once into avowed hostility. This view of our policy may, in truth, be extended to our connexion with every independent state in Asia. (II, 233)

## 2. From Lady Florentia Sale, *A Journal of the Disasters in Affghanistan, 1841-2* (London: John Murray, 1843)

[Born in Madras in 1787, Florentia Wynch was the daughter and granddaughter of East Indian Company civil servants. In 1808 she married Captain Robert Sale; he rose up the ranks of the British Army, and by the time that Florentia Sale arrived in Kabul in 1840, he was second-in-command and in possession of a knighthood (he was Henry Havelock's commanding officer). The journal covers the period from September 1841 to September 1842, and includes nine months of captivity at the hands of the Afghans. Florentia Sale died in 1853; her daughter and son-in-law, Alexandrina Sturt and Major Holmes, were among the very first British victims of the Indian "Mutiny," when they were beheaded by soldiers of the 12th Irregular Native Cavalry on 24 July 1857. *A Journal of the Disasters in Affghanistan* was a sales success, with four editions totalling over 7,500 copies being printed in its first year.]

*On the looting of the British cantonment in Kabul, 7 January 1842*

The reason the rear-guard were so late was, that they did not leave cantonments till sunset. Previous to their quitting them the Affghans had entered and set fire to all the public and private buildings, after plundering them of their contents. The whole of our valuable magazine was *looted* by the mob; and they burned the gun-carriages to procure the iron. Some fighting took place between the Affghans and our Sipahees. About fifty of the 54th were killed and wounded; and Cornet Hardyman, of the 5th cavalry, killed. A great deal of baggage and public property was abandoned in the cantonments or lost on the road; amongst which were two Horse Artillery six-pounders, as before mentioned. (228)

*Description of conditions during the retreat from Kabul, 11 January 1842*

We marched; being necessitated to leave all the servants that could not walk, the Sirdar promising that they should be fed. It would be impossible for me to describe the feelings with which we pursued our way through the dreadful scenes that awaited us. The road covered with awfully mangled bodies, all naked: fifty-eight Europeans were counted in the Tunghee and dip of the Nullah; the natives innumerable. Numbers of camp followers, still alive, frost-bitten and starving; some perfectly out of their senses and idiotic. Major Ewart, 54th, and Major Scott, 44th, were recognised as we passed then; with some others. The sight was dreadful; the smell of the blood sickening; and the corpses lay so thick it was impossible to look from them, as it required care to guide my horse so as not to tread upon the bodies: but it is unnecessary to dwell on such a distressing and revolting subject. (248-49)

*Captivity and the construction of British values, 10 May 1842*

What are *our* lives when compared with the honour of our country? Not that I am at all inclined to have my throat cut: on the contrary, I hope that I shall live to see the British flag once more triumphant in Affghanistan; and then I have no objection to the Ameer Dost Mahommed Khan being reinstated: only let us first show them that we can conquer them, and humble their treacherous chiefs in the dust. (342)

*Afghan insurgency and British collective punishment, 16 May 1842*

An European and some natives were murdered near our camp at Jellalabad; and vigorous measures not being taken, the offence was repeated; and a duffodar[1] of Tait's horse [guard] fell a victim to the Affghans. On the murderer taking refuge in a village, Tait immediately surrounded it with his men; and then reported the circumstance to Gen. Pollock; who, after consulting with Capt. Macgregor, sent to tell the people of the village that if they did not, within a specified time, give up the malefactor to be hanged by us, he would burn the village, and put every living being in it to death. The time had not expired when this news came. Cruel as an action of this kind may appear, it is probably the best method of striking terror into these savages, and perhaps of eventually preventing bloodshed. (348)

3. **From J.W. Kaye,** *History of the War in Afghanistan. From the Unpublished Letters and Journals of Political and Military Officers Employed in Afghanistan throughout the Entire Period of British Connexion with that Country* **(London: Richard Bentley, 1851)**

[Before he became the pre-eminent historian of the Indian "Mutiny," Kaye was the chief historian of the disastrous First Anglo-Afghan War. Kaye's *History* is trenchant and uncompromising in its criticism of the campaign, which he castigates on political, military, economic, and moral grounds. Kaye's assessment that the war had only stirred up

---

1 Persian, a non-commissioned cavalry officer.

resentment that would inevitably result in further conflict was widely shared; his explicit linking of the failure of the Afghan policy to the economic and political stability of India was prescient and astute.]

*Minutes of Sir Jasper Nicolls, 19 August 1841, testifying to the cost of the war*

My former proposal was met by an assertion (a very just one) that the heavy drain upon the finances would not admit an increase of establishment. I was not then aware of the full extent of that drain—it is now rated so high as to create a deficit of a million and a quarter annually, and I think we should not venture to send a second army beyond the Indus, to destroy the resources of India; for such a consequence may be apprehended from such a heavy annual exportation of the necessary funds. Again, when our jealous and intriguing neighbours observe our forces spreading to the east and west, so far beyond our former limits, and learn that our finances are decreasing annually, will they not be tempted to encourage each other to regain what we have wrestled from them, and to unite the turbulent spirits within our provinces to rebellion. (I, 654)

*Condemnation of the First Anglo-Afghan War as counterproductive*

It is on record, by the admission of Lord Auckland himself, that when our friendly connection with Afghanistan was brought suddenly to a violent and disastrous termination, it had cost the natives of India, whose stewards we are, more than eight millions of money. To this are to be added the cost of the great calamity itself, and the expenses of the War of Retribution. All this enormous burden fell upon the revenues of India; and the country is still groaning under the weight.

And what have we gained? What are the advantages to be summed up on the other side of the account? The expedition across the Indus was undertaken with the object of erecting in Afghanistan a barrier against encroachment from the West. The advance of the British army was designed to check the aggression of Persia on the Afghan frontier, to baffle Russian intrigues, by the substitution of a friendly for an unfriendly power, in the countries beyond the Indus. And now, after all this waste of blood and treasure, a Persian army is in Herat, and every town and village of Afghanistan is bristling with our enemies. Before the British army crossed the Indus, the English name was honoured in Afghanistan ... now in their place, are galling memories of the progress of a desolating army. The Afghans are an unforgiving race; and everywhere from Candahar to Caubul, and from Caubul to

Peshawur, are traces of the injuries we have inflicted upon the tribes. There is scarcely a family in the country which has not the blood of kindred to revenge upon the accursed Feringhees. The door of reconciliation is closed against us; and if the hostility of the Afghans be an element of weakness, it is certain that we have contrived to secure it ... the policy which we pursued was disastrous, because it was unjust. It was, in principal and in act, an unrighteous usurpation, and the curse of God was on it from the first. Our successes at the outset were a part of the curse. They lapped us in false security, and deluded us to our overthrow. This is the great lesson to be learnt from the contemplation of all the circumstances of the Afghan War. (II, 668-70)

## 4. From "The Murder of Lord Mayo," leader in *The Times*, 15 April 1872

[The Governor-General of India, Richard Southwell Bourke, the sixth Earl of Mayo (1822-72) was murdered by Shere Ali, an Afridi tribesman from Afghanistan, on the pier at Hope Town during his inspection of the penal settlement in Port Blair on 8 February 1872. Shere Ali had served the British as a mounted orderly in Peshawar and with the Punjab police from 1862; while in British service he murdered a relative in a long-standing blood feud in the city, and was sentenced to hang for the offence. In light of his earlier good character and service to the British, this sentence was reduced to life transportation to the new penal settlement at Port Blair on the Andaman Islands. Mayo's murder came just four months after the explicitly political assassination of Assistant Chief Justice Norman in Calcutta. Justice Norman had been responsible for suppressing a Pan-Indian Wahabi (Islamic revivalist) movement, directed largely by the chief imam of Patna, the Maulvi Ahmedullah, who was transported to the Andamans as punishment in 1864 (he ended up becoming the hospital clerk in Port Blair). British interpretation of the murder of Lord Mayo ranged from assertions of the innate and irredeemable barbarity of Imperial subject people (Shere Ali was seen as an archetype of native irrationality), to sophisticated speculations about new political conspiracies to undermine British rule. No evidence of a political conspiracy has ever been substantiated, and the most likely reason for Shere Ali's killing of Lord Mayo remains the most obvious one: that he was intent on taking revenge for what he felt to be a humiliating punishment for fulfilling the obligation of an honour killing. Nonetheless, Lord Mayo's murder unsettled British authority in India, and brought home once again, the uneasy relationship between Afghan policy, the penal settlement in the Andamans, and the prospects of another Indian uprising against the British (in this case, a specifically Islamic, Wahabi one).]

*Speculation about the political motivation for the assassination*

The plan for the murder appears to have been concocted long before the arrival of the Viceroy, whose intended visit was known here three month ago. A knife with which the deed was committed was prepared in the convict smithy (all Government knives are of different make and pattern). This could not have been done without the cognizance of the convict petty officers, who are nearly all Mussulmans, and the murderer, Shere Allee, openly declared that he would revenge the death of Abdoolla, Justice Norman's assassin ... whether the man was or was not a Wahabee, it is quite clear that he was of the material with which such men work. He had been born to a grievance as deadly as a Corsican *vendetta*. His family apparently had come to grief in that feud, and he had long before this, begun to look upon himself as the avenger. The fact that he complained of his sentence of transportation as unjust proves nothing, for he would look upon a murder in a blood feud as quite as much within the law of right as a sentence by the law of the land. That he thought himself an injured man is certain, and that feeling might and would exist, together with a full knowledge that he was a murderer. For the rest, he had free access to able, politic, daring Wahabee prisoners. Ahmedoola could not, if he had sought all India over, have found a better tool. It may be quite true that Shere Allee never in fact became a Wahabee. Sufficient that he could find men to remind him that the whole question of life and death could be compressed into a nut-shell, and that to kill a Feringhee was to secure the way to heaven [...] the execution was a triumph to the assassin. He went to his fate as to a feast, which would begin on Viper Island and end among the black-eyed houris of paradise.

# Appendix D: Colonial Contexts: The Andaman Islands

[First claimed for Britain by Lieutenant Archibald Blair in 1789, the first convicts were taken to the Andamans as unpaid labour later that year, and soon after, a small penal settlement was established in what is today Port Blair. The unhealthy climate and the hostility of the indigenous Jarawa led to the inevitable abandonment of this first penal settlement in 1796; the islands were effectively abandoned by the British for the next half-century. While the islands had been nominally under British sovereignty since Blair's 1788-89 expedition, it was the establishment of the penal settlement in March 1858—a direct response to the setbacks of the Indian "Mutiny"—that ushered in a new political order and significance for the Andaman Islands. No fewer than nine of the islands were immediately named after the fallen British heroes of 1857, including Havelock, Lawrence, Ross, Neill, and Outram. The settlement was stocked with both "mutineers" and common criminals, overwhelmingly from India, but with some prisoners from Burma as well. The punishment for escape from the penal settlement was death by hanging; in the first three months of the setting up of the penal colony, some 87 out of a total of 730 convicts had been hanged for desertion. At its peak in the 1890s, the penal colony held some 12,000 convicts. The penal colony in Port Blair became both an expression of punitive British colonial policy in India after the "Mutiny," and a visible reminder of the inefficacy of that policy. Indian nationalism was not suppressed through deportation to the Andamans; for the political prisoners, their time on the islands became a formative experience.

Despite 60 years of Indian Independence and a vogue for the decolonisation of place names, no attempt has been made to rename these islands, signifying their post-1947 political relevance as a reminder of the humiliation of transportation, and the desperation of the struggle for independence from British rule. The Andaman Islands have continued to be politicised in current historical discourse, with the Cellular Prison in Port Blair now firmly part of Indian nationalist mythology. The Andamanese, as ever, have been systematically elided from the historical debates over their homeland. The politicisation of the Cellular Prison by post-Independence Indian historians, novelists, and social commentators has recently been contested; see, for example, Satadru Sen's *Disciplining Punishment: Colonialism and Convict Society in the Andaman Islands* (New Delhi: Oxford UP, 2000).

The extracts below bring together journalism, anthropology, and official government reports. They provide a representative survey of British encounters with the Andamanese, the problems of pursuing the policy of penal settlement on the islands, and the progressive construction of an anthropological Andaman "archive," which in the twentieth century effectively replaced a decimated population with material artefacts, anthropometric photographs, relics, and academic studies. For further information about the fate of the Andamanese, now the most vulnerable population group in the world, see the suggested links in the bibliography.]

### 1. "The Andaman Islands, A Penal Settlement for India," letter to the editor of *The Times* (11 November 1857)

Sir,—I beg to offer a few suggestions which, in the present crisis of Indian affairs, may not be considered unworthy of your acceptance and consideration.

When the reinforcements so anxiously looked for in India shall at last have arrived from home, and we are enabled to follow up and punish those who have tortured and massacred so many hundreds of our countrymen, the question may arise, "How are we to dispose of so many felons?" Indiscriminate hanging, no doubt, would be the safest and easiest remedy; but our English notions of justice and our name as a Christian nation preclude all chance of "a consummation so devoutly wished for," and we shall therefore have on our hands a crowd of convicts, who either must be kept in prisons, to work in gangs on the roads in India, or be transported for life.

Now, even if the police of Bengal were to be depended on (which they are not), or the charge of convicts deemed the fitting occupation for English soldiers (which it is not), these home convicts would be both expensive and dangerous to the State. They would be objects of sympathy to all the disaffected in India, and their prisons the recruiting depots for any fresh enormity that might be attempted.

I have heard Ceylon, Mauritius, the Cape, New Zealand, and even the West India Islands recommended as places where they might be transported to; but apart from the heavy expense which such a proceeding would entail on the country, there are all Queen's colonies, and the inhabitants would, it is more than probable, object to such an importation.

Nearer India, and under the Company, we have Arracan, Burmah, and the Straits of Malacca, and convicts could be transported thither with ease and comparatively little expense, but to this also there are many objections.

These places are all too near India, and any convicts who might escape would soon find their way back to Bengal, burning to revenge themselves in a manner we can but too well imagine, to say nothing of the plots which the return of these men would encourage or give rise to.

Then it would be unwise to give these mutinous rascals a chance of infecting the Madras Sepoy, or of bragging to the natives, who collect in those parts from all quarters of India, that in them they should behold the patriots who endeavour to rid their country of Feringhees, and who thereby saved from pollution the very caste they still enjoy, but which the Feringhees had tried to deprive them of.

Within a few days' sail from Calcutta, and not very far from the coast of Burmah, washed on all sides by the Indian Ocean, lie the lands which I would suggest as the future penal settlement for India,— the Andaman Islands.

These islands are little known, save as the abode of savages, who have also the unamiable character of being cannibals and as a fatal place for shipwrecks, or at best an inexhaustible emporium for cocoanuts. Some of these islands are large, but, though belonging nominally to England, they have been turned hitherto to no account. Whether the climate or the natives, or both, have been the cause of this neglect on our part I know not, but such considerations ought not to be admitted for a moment in the case of the fiends whose guilt, if not actually proved by the laws of England, is known by every man in India to be dark and horrible, whose hands are stained with the blood of helpless women and children, and who are the basest and most treacherous of rebels, or, to say the least of them, who glory in the excesses and wickednesses of others, and connive at or allow the most horrible barbarities the world has ever known.

I think there is not a man in England who will not consider any place too bad for such men.

Let them be shipped off to the Andamans; let them be provided with a certain quantity of food and clothing, and with agricultural implements and seeds for their future sustenance, and then leave them to their fate. Defenceless as the women and children they have murdered, let them make head, if they can, against savages not more bloodthirsty than themselves; and the Bengalee, who was too proud and too mighty in his own eyes to touch the food over which the shadow of an Englishman had passed, will have to dig with his own hands before he can sow the rice that keeps his wretched body and soul together.

Should this be thought too severe a measure, then establish at once a footing on the Greater Andaman by means of the first batch of convicts, and make it the future penal settlement of India. Except those whose offences are comparatively trifling, and whose term of imprisonment is short, few convicts should be employed in future on the roads in India, for we have seen how in all these mutinies in Bengal to liberate the prisoners was the first thing attempted.

The high-caste Brahmins, or indeed, Bengalees in general, dread nothing so much as crossing the sea, or "dark water." It is pollution; and that is why few of them formerly enlisted for general service, and the fact of "general service" being now enforced on all fresh recruits is given as one reason for these outbreaks.

Therefore, let the worst felons be shipped beyond the seas, and the prospect of Andaman will be worse than death itself to the "gentle Hindoo," whose caste, undefiled by innocent blood, can still, happily, be washed away by water. Surely for such awful crimes the punishment is light; and if we may not cast human beings away to perish unprotected, let them clear ground and build a fort for the soldiers who guard them, and the lighthouse which is so much needed in these seas and on these very islands; and instead of Andaman being the dread of every native craft in India, as it is now, its naturally good harbour will become a haven of safety for the ships of all nations, and in time a flourishing colony will spring up where, for want of clearing and cultivation, all is barren and waste.

I might go further. When these wretched mutineers shall find themselves entirely cut off from their native land, their caste lost to them for ever, and themselves dependent for the means of life and defence on the very race they have tried to exterminate, if any feelings of gratitude or remorse can exist in such hearts it will be awakened, and some few hardened sinners may yet be brought to acknowledge their sin and their Saviour before they pass from earth to Him whose dread instruments of punishment they have been.

One more observation concerning our occupation of the Andaman Islands I wish to make before concluding this letter. Might it not be a judicious and politic move on our part? For should England be again engaged in another great war, these islands are too near India not to escape (if unoccupied) the attention of her enemy, and a hostile fleet, with the shelter and harbourage, provisions and water, which the Andamans afford, would be an awkward neighbour for our Indian Empire.

## 2. From Frederic J. Mouat, *Adventures and Researches Among the Andaman Islanders* (London: Hurst and Blackett, 1863)

[Frederic John Mouat (1816-97) was a surgeon-major in the British Indian Army, and an inspector general of prisons in Bengal. Following his participation in the successful establishment of the penal colony at Port Blair (he was President of the Committee and in general charge of the expedition of December 1857), Mouat published one of the first accounts of the British encounters with the Andamanese. One of Mouat's more remarkable acts was the capture of a young Andamanese man (dubbed Jack) who was subsequently presented to the Governor-General in Calcutta, Lord Canning; housed by Mouat, "Jack" became a Calcutta spectacle. Mouat's *Adventures and Researches* was later challenged by E.H. Man, and even more comprehensively by Maurice Vidal Portman, for being factually inaccurate and misleading. Portman dismissed Mouat's book for containing "numerous" errors, based on a limited experience and knowledge of the Andamanese: "Dr Mouat came into contact with the Andamanese on four occasions only, each of a few hours' duration. He knew nothing of their language, and the meetings were hostile on the part of the savages" (Portman, *A History of our Relations with the Andamanese*, 210). Today, Mouat's account's is interesting more for what it says about the author than about his putative subjects. Reeking with the author's assumptions of racial, cultural, and moral supremacy, Mouat's *Adventures and Researches* provides a personal insight into high Victorian imperialism; he explicitly links the fate of the islands with the events of the summer of 1857.]

*On the "savagery" of the Andamanese*

In these islands man is probably still in the same condition in which he was when they were first settled—a state not only simple and primitive, but lower even than that of the natives of those remote islands of the sea that lie far beyond the bounds of civilization. Their inhabitants acknowledge no law to restrain and guide them in life. Not only have they no knowledge of a supreme Being, but they are even destitute of such a rude system of religious faith as is generally found even amongst men in the most savage state. Their language is harsh and disagreeable, in an imperfect state of development, barely sufficient for the expression of their few and simple wants. Totally ignorant of agriculture, they have no means of making provision for the future, and must be satisfied with such fruits and herbs as the earth spontaneously produces, or with the shell-fish they pick up on the shore. Their habitations are of that primitive and simple construction which suffices for the necessities of savage tribes constantly moving from place to place ... the inhabitants of the Andamans have always been considered one of the most savage races on the face of the earth, whom civilization has yet found it impossible to tame, or even almost to approach. From the earliest records we have, down to the more recent descriptions of their appearance and character, all accounts agree in this unfavourable reputation, and the opportunities of obtaining an accurate knowledge of them having been hitherto so few and unfavourable, much has never been known with regard to their actual condition, their habits and modes of life, or their origin and peculiar characteristics. The study of the race to which they belong also involves one of those ethnological problems which yet remain to be solved; and although it may not at any rate be in our power, by means of the personal observations we have an opportunity and making, to furnish some data by which a solution may be rendered easier to other inquirers. (2-4)

*On the rationale behind the establishment of the penal settlement*

The object in view was not only to make the islands safe asylums for those who had the misfortune to be wrecked on their coasts, but also to utilize them in such a way as would prove ultimately beneficial to the inhabitants themselves, supposing their suspicious fears overcome, and their confidence gained. Two plans were accordingly proposed. One was the formation of a harbour of refuge on a suitable part of the coast, the expediency of which was generally admitted. The second was the establishment of a penal settlement on the principal island, in the most advantageous locality that could be selected. The advisability of carrying this proposal into effect was under discussion in the Indian Council at the time when the late dreadful mutiny broke out in 1857. Startling the world by its sudden, savage, and unprovoked nature, its immediate effect was to lay all such useful measures and plans in abeyance for the time; and it was not until the neck of treacherous rebellion was completely broken that the subject was again submitted to the Councils of the Government, in circumstances that rendered a speedy decision necessary. A settlement was now required to which those misguided agents of the late mutiny, whose crime, however great, was not attended with circumstances of such unpardonable atrocity as rendered imperative the forfeiture of their lives, might be transported. There were many whose hands had not been actually imbued in blood, yet who, from the share they had openly taken in the revolt, could not with safety be included in any measure of amnesty, however comprehensive, until either the last traces of disaffection had entirely disappeared, or the natives of India were thoroughly convinced that any further attempt at rebellion against the authority of England must infallibly be put down. It was believed that the transportation of these mutineers to the Andaman Islands would be an adequate punishment for the crime of which they had been guilty. There was something poetical in the retributive justice that thus rendered the crimes of an ancient race the means of reclaiming a fair and fertile tract of land from the neglect, the barbarity, and the atrocities of a more primitive, but scarcely less cruel and vindictive race, whose origin is yet involved in such a dark cloud of mystery. (44-45)

*On the appearance of the Andamanese*

We had picked up two of the bodies of the three Mincopie[1] who were killed during the fight, and in obedience of the orders I have previously given, they were now laid out on deck, and decently covered with matting ... the two men were, like all of their race whom we had seen, of short stature, but their conformation appeared to be remarkably robust and vigorous, and the muscular development of their arms, legs and chest was very considerable. Their countenances were anything but agreeable, pleasing, or attractive. Their expression, as it had been settled by the hand of death, had something in it that was truly repulsive and frightful. Their features, distorted as they appeared by the most violent passions, were too horrible for anything of human mould, and I could regard them only as the types of the most ferocious and relentless fiends. If the whole race resembles these two men, we had no reason to be surprised that we had failed in every attempt to arouse friendly and hospitable feelings in a tribe whose aspect was really that of demons. (255-56)

*Exhibiting "Jack"*

When the news spread in the city that a native of the Andamans had been brought back with the expedition, it was astonishing with what rapidity the report, which, at such an exciting period, one would have considered of little importance, spread from mouth to mouth, creating an amount of excitement we had never ventured to anticipate. Jack became a sort of nine days' wonder, and in their intense desire to see him, the inhabitants, both native and foreign, came crowding in hundreds in front of my house ... the stories, too, that were circulated regarding Jack, who had now assumed a very quiet and composed air, were such as demanded a considerable amount of faith, a commodity which is always ready for any absurdity. That Jack was a cannibal was a point no one ventured to deny, while the common reports about his teeth, his feet, and other parts of his body, were so monstrous that I will not violate probability by stating them ... the absurdity of the exaggerated stories that were everywhere circulated was such that it would be almost equally absurd to repeat them. They were the means, however, of keeping alive the general curiosity to see so rare a monster

---

1  An earlier term for the Andamanese, coined by Colebrooke in the 1794 expedition to the islands and by this time already archaic.

as they assumed poor good-natured Jack to be, and the crowds increased in number every day and every hour, until the whole neighbourhood was in a state of excitement. (280-82)

*Refutes the charge of cannibalism*

But they are certainly not so bad as they have been described. The statement that they are cannibals is unworthy of belief, for it rests on no trustworthy foundation. No one ever saw them indulging in those horrid banquets which have been attributed to them. Any case of their actually devouring human bodies is unknown, and we have no reason to suppose that they even devour raw flesh of any kind. (302)

3. **From the *Annual Report on the Settlement of Port Blair and the Nicobars for the Year 1872-3* (Calcutta: Office of the Superintendent of Government Printing, 1873)**

[These annual reports provide a comprehensive overview of the administration and conditions of the penal settlement on the Andaman Islands. For example, the report tells us about the number of prisoners in the penal colony, as well as their classifications, and the size of the military garrison housed on the islands (118 British and 301 Indian soldiers). The assassination of Lord Mayo on 8 February 1872 brought renewed scrutiny to the administration of the Andamans in general, and the penal colony in particular.]

*The anomalous nature of the penal settlement*

It seems desirable here to record that the system of convict management now in force is the growth of circumstances. And bears little resemblance to the penal settlements established in the British Colonies and elsewhere, and all things considered, it is creditable to the system and to those who worked it that it did not collapse altogether. (1)

*The system of discipline and the policy of internal policing outlined*

The Northern Stations are Perseverance Point, Hope Town and Mount Harriet, with the sub-station of Bamboo Flat.

On all these stations the convicts are quartered in open barracks very much like soldiers in a military cantonment, and the whole of the interior economy and discipline are carried on by petty officers selected from the body of the convicts under the supervision of the

European officer in charge of the stations, aided in most, though not in all, cases by a free European overseer.

For the preservation of order outside the barracks, there is a body of convict police, and at all the chief stations there are small detachments of free police which provide guards for the officers, store-godowns, lock-ups, etc., but which have no direct control over the convicts.

It will thus be seen that the convicts are practically managed by officials selected from their own body, and that the arrangement is a wise one, for it encourages good conduct and intelligence, and provides desirable employment for a considerable number of the criminal population, whose interests are thereby enlisted on the side of the authorities. (3-4)

*The frequency of escapes*

The escapes have been so numerous during the year 1872 that the subject calls for something more than a mere numerical statement of facts.

They amount in all to 160, and of these 100 escaped by sea.

Of the 60 convicts who escaped by land, 44 were re-captured during the year, and 49 of those who escaped by sea were likewise re-taken, one having died at sea.

Escapes by land are no longer of much importance, because the convicts are alive to the fact that without a boat of some sort there is no escape from the islands, and that the only alternatives are a miserable lingering death from exposure and starvation, and recapture.

On the other hand, the numerous escapes by sea point to a defect in the economy of the settlement, for which it is difficult to find an effectual remedy, as the most careful precautions yet devised have been powerless to guard against the opportunities to escape to which the convicts employed in boats are unavoidably exposed.

So long as the boats were manned entirely by term convicts, escape by sea was not of frequent occurrence; but these are no longer available, and of the 460 men regularly employed in boat-work, more than three-fourths are life convicts, and their numbers are daily increasing. (6)

*Size and description of the convict population*

The average number of prisoners during the year has been 6,506 males and 733 females, total 7,239 ... in consequence of the large number of term-prisoners whose sentence has expired the past 2 years, the population of prisoners has diminished, but, as the number of term-prisoners are now becoming very small, a gradual increase in the number of convicts may be looked for ... during the past year the number of European and East Indian prisoners have decreased, either by casualties, releases or other causes [the table lists 22 "Christians" amongst the convict population, but does not say whether they are Indian or British] (54-55).

### 4. From "The Andamans Penal Settlement," *The Times* (13 February 1872)

To judge from what has lately transpired concerning the usual life of a convict in the Andaman Islands, the penal settlement there has for some time past been turned into a paradise of rum-drinking and unlimited idleness. It appears that the European and Eurasian convicts, at any rate, have been allowed to do pretty much as they pleased. They go freely into each other's rooms, wander where they like outside, take into their service the sepoys who are supposed to guard them, entertain their friends to dinner, and are free to draw for a whole gallon of rum at one time. Unluckily the dinners and the drinking sometimes lead to quarrels, which now and then lead to bloodshedding or downright murder. All this came out at the Calcutta Criminal Sessions on May 6th, when a Port Blair convict, James Devine, was convicted of murdering a comrade in a drunken quarrel, the sad but not unnatural close of an evening spent by Devine and his friends in getting through a gallon of rum. Devine, being mad drunk, battered in the head of the man who had lain nearest him that night. He was found guilty, but recommended to mercy on the plea that "the crime would not in all probability have been committed but for the *disgraceful laxity of discipline* and want of proper control over the convicts at Port Blair, as shown in the evidence." Whatever becomes of this particular ruffian, we may hope that General D. Stewart, the new Governor of the Andamans, will remove like temptations to like deeds of violence out of the convicts' way. Lord Napier, we are told, has long sighed for a little more discipline at Port Blair, and General Stewart, as being his own selection, may be trusted to carry out the desired reforms.

### 5. From "The Andaman Settlements: From Our Own Correspondent," *The Times* (26 December 1873)

*On the regime of the penal settlement on Viper Island*

The chain gang consisted of some 200 ruffians, heavily ironed, and all paraded, with their bedding and utensils, for the apothecary's inspec-.

tion. Every one was a murderer or a mutineer, or a convict who had deliberately and frequently broken the prison rules [...] most of the solitary cells contained an occupant, and each cell is larger and airier than any I have seen on the Continent of India. There was little in this place dissimilar from what may be seen in the large gaols of England, except the arrangements for eating.

## Penal settlement modelled on Agra jail

On the 4th of March, 1858, the first batch of 200 convicts was landed to clear Blair's old settlement on Chatham Island. Dr. J.P. Walker was selected as the first Superintendent, because of the great success with which he had met in the treatment of the worst criminals in what he had made the model gaol of Agra.

## On the system of self-policing the penal settlement

There are a very few free overseers and clerks, but all the petty officers are convicts who have earnt that position by good behaviour. This is an important feature in the system, without which it could not be carried out save at enormous expense. It works admirably so long as there are short-term prisoners from which to select men whose interest lies in doing their duty. But gradually as this class is being released—and there will soon be only life-convicts, rendered desperate by the absence of hope—there will be no such petty officers, and I do not see how the settlement can exist.

## On the "ticket-of-leave" system and manumission

Each convict on arrival is placed in the third class after due inspection. All his antecedents are well known, but for Settlement purposes, ignored. He begins a new life, and on his character and acts in the Settlement alone depends his treatment. He must serve in this class for seven years, at hard labour from 6 to 11 and 2 to 6, having as little intercourse with his fellows as possible, and receiving only cooked food. If he passes through this period without having been placed in the chain gang or refractory ward at Viper Island, he enters on five years' experience of the second-class. His work is the same, but he may marry or send to India for his wife and family, may cook his own food, may make earnings by extra work, and may after one year be appointed a petty officer. In the first class he is paid more, may be promoted to the superior grades of petty officers, and may be apprenticed to private employers, or get a Government situation. In the case of marked good conduct, the life-convict may get a ticket-of-leave after 12 years, becoming a self-supporter and a producer for the good of the Settlement as well as of his own pocket. This gives him freedom within any part of the Colony, but he must report himself every half-year. He may accumulate personal property, but not land. Finally, a conditional pardon, after 21 years, is held up before the life-convict, who may, in special cases, be allowed to leave the place, but if not, becomes a free man in the Settlement, subject only to the duty of reporting himself on the first day of every year.

*On the absence of British convicts*

There are no English convicts now at Port Blair. The last, a Scotch seaman who knocked down a subordinate in a moment of passion, has served his time, and remains a free overseer of the station where he has so long worked as a convict [...] a hundred romances might be written from the histories and the correspondence of these and other convicts.

6. **From the *Annual Report on the Settlement of Port Blair and the Nicobars for the Year 1873-4* (Calcutta: Office of the Superintendent of Government Printing, 1874)**

[Responding to increasing alarm about the lawlessness of the settlement, this annual report is keen to emphasise the establishment of a freshly recruited "free" police force to control the convict population.]

*Size and make-up of the new "free" police force*

The free police establishment of this settlement consisted, on the 31st December 1873, of 330 men of all ranks ... during the year 225 men were enlisted in the force; of these, 219 were entertained in the North-West Provinces and the Punjab. (17)

*Size and make-up of the European population, extracted from Section XIII*

Europeans and their descendants:
Civil, males: 45; Military, males: 108; Marine, males: 3; Police, males: 2; Free residents, females: 33; Children of free residents, males/females: 32; 21; Convicts, male: 6. (36)

7. **From Edward Horace Man, *On the Aboriginal Inhabitants of the Andaman Islands* (London: Published for the Anthropological Institute of Great Britain and Ireland by Trübner & Co., 1884)**

[Previously superintendent of convicts in Moulmein in Lower Burma, Colonel Edward Horace Man (1846-1929) lived in the Andaman Islands between 1869 and 1880, and again from 1882 to 1889. He was responsible for running the "homes" established by the administration for the "pacification" of the Andamanese. The homes were designed to bring the Andamanese out of the forest, facilitate contact with them, and prevent them from attacking the penal settlement. Despite their seemingly benign intent, the homes policy proved disastrous, for it spread imported diseases (such as syphilis and measles) to which the Andamanese had no resistance. Man translated parts of the Bible— *The Lord's prayer, translated into the Bojingijida, or South Andaman (Elakabeäda) language* (Calcutta: Thacker, Spink & Co., 1877)—and with Richard Carnac Temple, wrote the first grammar of an Andamanese language, *Grammar of the Bojingijida or South Andaman language* (Calcutta: Thacker, Spink & Co., 1878). Man's work on the islands is a representative example of the parochialism of much late nineteenth-century British anthropology and colonial policy; without recourse to any systematic methodology, he provided the first detailed account of the Andamanese, while also simultaneously presiding over a catastrophic decimation of their numbers caused by the very policy that he espoused.]

*On the stature and appearance of the Andamanese*

From my own observations I would remark, that though it is quite true that there are found among them individuals whose "abdomens are protuberant, and whose limbs are disproportionately slender," such persons no more represent the general type of the race, than the sickly inmates of a London hospital can be regarded as fair specimens of the average Englishman ... the average height of the men is 4 feet 10¾ inches, and that of the women 4 feet 7¼ inches ... the old statement, so often repeated, that their stature never exceeds 5 feet, must also be noticed, as the list in question shows that fourteen of the forty-eight males who were measured were slightly above that height. (4-6)

*Refutes the myth of cannibalism*

The early stories regarding the prevalence of cannibalism among these savages do not at the present day require refutation. No lengthened investigation was needed to disprove the long credited fiction, for not a trace could be discovered of the existence of such a practice in their midst, even in far-off times. (45)

*On the Andamanese use of stone tools*

Although a great portion of the inhabitants of Great Andaman have for some time past been able to procure iron in sufficient quantities to substitute it for stone (not to mention bones and shells), still they can by no means be said to have passed out of the stone age; indeed, the more distant tribes still retain the use, with scarcely any modification, of most of the stone and other implements which served their ancestors ... the whetstone and hammer only are offered as mediums of exchange, but no great value is attached to either of these objects, nor is any superstition associated with their usage; they are therefore—when no longer serviceable—cast aside ... the Andamanese, however, maintain that they never, even when iron was scarce, made arrowheads, axes, adzes, or chisels of stone. (159-61)

*On the Andamanese ignorance of the blow-pipe and poison*

The blow-pipe, which is used so generally by the Negritos (*Semangs*) of the Malayan peninsula, finds no place amongst the weapons of these savages. Its absence may be readily accounted for, firstly by their ignorance of poison, or at least any method of utilising such knowledge as any of them may possess, and secondly by the fact that they are so well able to supply all their wants with the implements already referred to in the foregoing, that their inventive faculty has not been sharpened by the pangs of hunger to devise other or more effective means of destruction. (142)

## 8. From Maurice Vidal Portman, *A History of Our Relations with the Andamanese. Compiled from Histories and Travels, and from the Records of the Government of India* (Calcutta: Office of the Superintendent of Government Printing, 1899)

[Entering the civil service as an officer in the Indian Government Marine, Maurice Vidal Portman (1861-1935) was appointed "officer in charge of the Andamanese" in July 1879, and remained resident on the islands almost without interruption until his retirement from the civil administration in 1901. Portman's reputation as the leading nineteenth-century expert on the Andamanese has stood the test of time, despite pointed criticism from later professional anthropologists, such as Radcliffe-Brown. As well as the two volume *History*, he wrote *A Manual of the Andamanese Languages* (London: W.H. Allen & Co., 1887), and *Notes on the Languages of the South Andaman group of tribes* (Calcutta: Office of the Superintendent of Government Printing, 1898), but his most lasting contribution to Andaman scholarship remains the remarkable, and at times disturbing, photographic collection of nearly 300 items amassed during his residence in the islands. Some of the photographs clinically record the contemporary scientific practice of anthropometry (racial classification through physical measurement), while others are remarkably empathetic (and even elegiac) portraits of individual Andamanese. Portman's photographic collection is the largest of its kind ever assembled, and traces the effective effacement of a people by the very archive used to define, record, and govern them. For an assessment of the importance of Portman's images, see John Falconer, "Ethnographical Photography in India, 1850-1900," *Photographic Collector* 5.1 (1984): 16-46.]

*Exhibiting a group of Andamanese in Calcutta*

In January 1884, I was directed to proceed to Calcutta in charge of 37 Andamanese, including two women, and five Nicobarese, in order to show them at the exhibition held there.

The Andamanese taken by me were selected from the different tribes, and chosen on account of their superior influence, position, and intelligence. In addition to the Exhibition they were shown many things which I thought would be likely to interest them and attracted considerable attention. Perhaps the circus was what they appreciated most, and after that the Zoological Gardens, and the Museum, where they quickly recognised and named specimens of fishes, etc. (659)

*Assessment of the population for the annual report, 1890-1*

All the people on Rutland Island and Port Campbell are dead, and very few remain in the South Andaman and the Archipelago. The children do not survive in the very few births which do occur, and the present generation may be considered as the last of the aborigines of the Great Andaman. Even these have their constitutions to a great extent undermined by hereditary syphilis, and are unable to endure much exposure.

I am endeavouring to keep the race alive as long as possible, and am collecting all the children at my house, where they are well fed and looked after, and are taught to pull in my boat, wait at table, and make themselves generally useful. I have also provided a gymnasium for them. I have now 36 boys with me. (676)

*On the failure of the homes and the extinction of the race*

I have been frequently asked, "Why are the Andamanese race dying out?" The reason has generally been supposed to be, "on account of the syphilis introduced among them by the convicts," but I do not entirely agree with this ... the Andamanese have had no fresh blood for many centuries, and continued in-breeding has weakened their constitutions. The savage, far from being, as people so often suppose, a robust man, is generally very delicate. He can do certain surprising feats in the water or on land because he is accustomed to do them, but he cannot compare with a European in his endurance of new hardships and altered circumstances. Had the Andamanese been left entirely alone, no doubt they would have continued to exist for many centuries in the same state, and it is possible that the Öngés of the Little Andaman, if so left alone, may do so, and as we require nothing from them except that they should be friendly, and help instead of massacring the crews of ship wrecked vessels, care should be taken that this tribe is not interfered with more than is absolutely necessary; but with the aborigines of the Great Andaman circumstances were different.

It was found necessary, on our occupation of the Islands in 1858, to prevent the Andamanese from opposing the development of the Settlement, from murdering the convicts, and later on, from plundering.

They were then utilised in capturing runaway convicts, and to compensate them for the annexation of their country, and to cement and continue the friendly relations with them. Homes for them to which they were encouraged to resort, were established with the best intentions. These Homes, however, were most deleterious, in them the Andamanese learnt to smoke, contracted new diseases, and were given new foods to which they were unaccustomed. Their customs and modes of life were also altered; several well-meaning but mistaken persons were anxious that they should change their way of life entirely, and should settle down to agriculture ... so long as they were left to themselves and not in any way interfered with by outside influences, or their customs, food, etc, altered, they would continue to live; but when we came amongst them and admitted the air of the outside world, with consequent changes, to suit our necessities, not theirs, they lost their vitality, which was wholly dependent on being untouched, and the end of the race came. (873-75)

# Appendix E: Contemporary Reviews

[The volume publication of *The Sign of Four* was not reviewed particularly widely or well, perhaps because its appearance in *Lippincott's Monthly Magazine* had already been noted earlier in the year, but also because its publication was during the crowded pre-Christmas season for book releases. As a result, many reviews combined a selection of recently released novels (see reviews 4, 5, and 6), and often gave Doyle's second Sherlock Holmes story relatively short shrift. However, several common themes are evident in the reviews: general admiration of the narrative thrust of the detective tale (and especially the chase on the Thames); comparison with Wilkie Collins's *The Moonstone* (1868); reservations about Jonathan Small's narrative; and recognition of the singular novelty of Holmes. The two South African reviews (numbers 7 and 8) demonstrate the reach of Doyle's novella in the context of colonial publishing and reading, while Andrew Lang's essay on Doyle's novels (extracted here) constitutes the first full literary response to *The Sign of Four*, and is notable for both its precise criticism of Doyle's carelessness of significant detail (especially with regard to the Andamans) and its appreciation of the detective novel as an entertaining and influential literary genre.]

1. **Anon., "Magazines for February,"** *Liverpool Mercury*, **5 February 1890, p. 7**

*Lippincott's* (Ward, Lock and Co., London) has an inviting variety of contents. Its complete story this month is by A. Conan Doyle, and has for its title "The Sign of the Four." The story brims with incident, and the interest is sustained from the first chapter to the close.

2. **Anon., "Notes on Novels,"** *Dublin Review*, **April 1890, pp. 442-43**

Mr. Conan Doyle's novelette is a tale of an Indian treasure, and bears traces of books that have already been written—such as "The Moonstone." It contains a wonderful detective, an Andaman islander, and a very good chase on the Thames. The weak point of the story is when the villain, being caught, relates in a style too much like Mr. Doyle's own, how "he came to" get hold of the treasure. As this occurs at the very end, when the Andaman Islander has been shot, the box ("of Indian workmanship") found to be perfectly empty, and the preternatural detective again become more or less comatose, the result is not good art.

### 3. Anon., "Novels of the Week," *The Athenaeum*, 6 December 1890, p. 773

A detective story is usually lively reading, but we cannot pretend to think that "The Sign of Four" is up to the level of the writer's best work. It is a curious medley, and full of horrors; and surely those who play at hide and seek with the fatal treasure are a strange company. The wooden-legged convict and his fiendish misshapen little mate, the ghastly twins, the genial prizefighters, the detectives wise and foolish, and the gentle girl whose lover tells the tale, twist in and out together in a mazy dance, culminating in that mad and terrible rush down the river which ends the mystery and the treasure. Dr. Doyle's admirers will read the little volume through eagerly enough, but they will hardly care to take it up again.

### 4. Anon., "New Novels," *The Academy*, 13 December 1890, p. 561

Detective stories have always had a certain charm, and perhaps the charm is greatest when the detective element is non-professional. The accomplished amateur in the fine art of discovering crime and hunting down the criminal is a much more wonderful personage than the official detective. At any rate, Sherlock Holmes, in *The Sign of Four*, was such a personage. The curious incidents, the mystery of which he unravels, make a capital story, which is told with a directness that keeps the reader's attention fixed till he gets to the sequel. After the sequel, as part of the story, follows the narrative of the man who has been hunted down; and though this is interesting in itself, and has a bearing upon the plot, it is somewhat flat after a breathless chase which has been breathlessly described. It has the effect of an anti-climax. Sherlock Holmes is the best drawn of the characters, perhaps because there was most character in him to draw. The young lady is rather insipid, but she has not much to say or do. The man with a wooden leg, who was nearly a match for Holmes, is also nearest to him in point of portraiture.

### 5. Anon., "A Batch of Novels," *Liverpool Mercury*, 24 December 1890, p. 7

In Mr. Doyle's "The Sign of Four" ... there is no disguising the fact that we are engaged in following the story of crime and its detection pure and unmitigated. The records of amateur detective work undoubtedly have a fascination of their own; and Mr. Doyle is inexhaustible in the devising of trains and coils of circumstance which shall throw into brilliant relief the preternatural sagacity of his amateur detective. He tells us, indeed, that the detective's medical friend falls in love at sight with the heiress whose treasure has been conjured away, and we duly learn at the end that the pair were blissfully wedded. But such trifles are only introduced to save the character of the book as a "novel." The writer's interest, and the reader's too, if he has any, are throughout absorbed in tracking the guilty and deciphering the mysterious "Sign of Four!"

**6. Anon., "New Novels," *The Graphic*, 7 February 1891, p. 150**

"The Sign of Four," by A. Conan Doyle (1 vol.: Spencer Blackett), is a romance of the Detective school, in which the chief figure will be recognised by readers of "A Study in Scarlet" in the person of an amateur who must be either engaged in unravelling first-class mystery or in consoling himself for the want of one with cocaine. The mystery in the present case is very decidedly of the first-class, demanding all his genius and energy, and that he rises to the occasion is no less certain. It is not even any professional detective who could recognise, from certain small signs, that a mystery must be somehow mixed up with a native of the Andaman Islands, whose peculiarities will prove new even to the habitual novel reader, who has now come across most things. The story—which faintly suggest the machinery of the "Moonstone"—is entertaining in a rough and crude way, and its "creepiness" leaves little to be desired even by the most exacting.

**7. Anon., "Review of Books," *The Cape Illustrated Magazine* (Cape Town), 1 October 1894, p. 71**

The usual ingeniously constructed mystery is here unravelled by the wonderful amateur detective, for the edification of his friend, who narrates it to his readers in his well known perspicuous style. One little matter that goes a long way towards making this edition[1] popular, is the good paper, and very readable type employed.

---

1 Longman's Colonial Library edition, supplied by Messrs Juta & Co. of Cape Town.

## 8. Anon., *The Cape Illustrated Magazine* (Cape Town), 1 November 1894, p. 105

A ghastly and apparently impenetrable mystery is here unravelled in the author's well-known style. Those who are inclined to be critical, may accuse him of having dropped into a "groove" from which it seems impossible for him to escape, but that very fact has also its flattering side as there is no doubt that a steady demand for his work influences the supply. As a matter of fact there is always a welcome for detective stories, and those of Mr. Doyle are greatly in advance of anything else of the kind.

## 9. From Andrew Lang, "The Novels of Sir Arthur Conan Doyle," *The Quarterly Review*, July 1904, pp. 158-79

It may be a vulgar taste, but we decidedly prefer the adventures of Dr. Watson with Mr. Sherlock Holmes. Watson is indeed a creation; his loyalty to his great friend, his extreme simplicity of character, his tranquil endurance of taunt and insult, make him a rival of James Boswell, Esq., of Auchinleck. Dazzled by the brilliance of Sherlock, who doses himself with cocaine and is amateur champion of the middle-weights, or very nearly (what would the Bustler's trainer say to this?), the public overlooks the monumental qualities of Dr. Watson. He, too, had his love affair in *The Sign of Four*; but Mrs. Watson, probably, was felt to be rather in the way when heroic adventures were afoot. After Sherlock returned to life—for he certainly died, if the artist has correctly represented his struggle with Professor Moriarty—Mrs. Watson faded from this mortal scene.

The idea of Sherlock is the idea of Zadig in Voltaire's *conte*, and of d'Artagnan exploring the duel in "Le Vicomte de Bragelonne," and of Poe's Dupin, and of Monsieur Lecoq; but Sir Arthur handles the theme with ingenuity always fresh and fertile; we may constantly count on him to mystify and amuse us. In we forget what state trial of the eighteenth century, probably the affair of Elizabeth Canning, a witness gave evidence that some one had come from the country. He was asked how he knew, and said that there was country mud on the man's clothes, not London mud, which is black. That witness possessed the secret of Sherlock; he observed, and remembered, and drew inferences, yet he was not a professional thief-taker.

The feats of Sherlock Holmes do not lend themselves as inspiring topics to criticism. If we are puzzled and amused we get as much as we want, and, unless our culture is very precious, we *are* puzzled and amused. The *roman policier*[1] is not the roof and crown of the art of fiction, and we do not rate Sherlock Holmes among the masterpieces of the human intelligence; but many persons of note, like Bismarck and Moltke, are known to have been fond of Gaboriau's tales.[2] In these, to be sure, there really is a good deal of character of a sort; and there are some entertaining scoundrels and pleasant irony in the detective novels of Xavier de Montépin and Fortuné du Boisgobey,[3] sonorous names that might have been borne by crusaders! But the adventures of Sherlock are too brief to permit much study of character. The thing becomes a formula, and we can imagine little variation, unless Sherlock falls in love, or Watson detects him in blackmailing a bishop. This moral error might plausibly be set down to that overindulgence in cocaine which never interferes with Sherlock's physical training or intellectual acuteness ... Edgar Allan Poe, who, in his carelessly prodigal fashion, threw out the seeds from which so many of our present forms of literature have sprung, was the father of the detective tale, and covered its limits so completely that I fail to see how his followers can find any fresh ground which they can confidently call their own. For the secret of the thinness and also of the intensity of the detective story is that the writer is left with only one quality, that of intellectual acuteness, with which to endow his hero. Everything else is outside the picture and weakens the effect. The problem and its solution must form the theme, and the character-drawing be limited and subordinate. On this narrow path the writer must walk, and he sees the footmarks of Poe always in front of him. He is happy if he ever finds the means of breaking away and striking out on some little sidetrack of his own.

Not much more is left to be said by the most captious reviewer. A novelist writes to please; and if his work pleases, as it undeniably does, a great number and variety of his fellow-citizens, why should his literary conscience reject it? If Poe had written more stories about Dupin—his Sherlock Holmes—and not so many about corpses and people buried alive, he would be a more agreeable author. It is a fact that the great majority of Sherlock's admirers probably never heard of Poe; do not know that detective stories date from Dupin ... possibly the homicidal ape in "The Murders in the Rue Morgue" suggested the

---

1 Detective novel (French).
2 French novelist Émile Gaboriau (1832-73) created the fictional detective Monsieur Lecoq, an influence on Sherlock Holmes.
3 Xavier de Montépin (1823-1902), bestselling French novelist of the 1880s with the *La Porteuse de pain* (1884-87) series. Fortuné du Boisgobey (1821-91) wrote popular French detective stories.

homicidal Andaman islander in *The Sign of Four*. This purely fictitious little monster enables us to detect the great detective and expose the superficial character of his knowledge and methods. The Andamanese are cruelly libelled, and have neither the malignant qualities, nor the heads like mops, nor the weapons, nor the customs, with which they are credited by Sherlock. He has detected the wrong savage, and injured the character of an amiable people. The *bo:jig-ngijji* is really a religious, kindly creature, has a Deluge and a Creation myth, and shaves his head, not possessing scissors. Sherlock confessedly took his knowledge of the *bo:jig-ngijji* from "a gazetteer," which is full of non-sense. "The average height is below four feet!" The average height is four feet ten inches and a half. The gazetteer says that "massacres are invariably concluded by a cannibal feast." Mr. E.H. Man, who knows the people thoroughly, says "no lengthened investigation was needed to disprove this long-credited fiction, for not a trace could be discovered of the existence of such a practice in their midst, even in far-off times."

In short, if Mr. Sherlock Holmes, instead of turning up a common work of reference, had merely glanced at the photographs of Andamanese, trim, elegant, closely-shaven men, and at a few pages in Mr. Man's account of them in "The Journal of the Anthropological Institute" for 1881, he would have sought elsewhere for his little savage villain with the blow-pipe. A Fuegian who had lived a good deal on the Amazon might have served his turn.

A man like Sherlock, who wrote a monograph on over a hundred varieties of tobacco-ash, ought not to have been gulled by a gazetteer. Sherlock's Andamanese fights with a blow-pipe and poisoned arrows. Neither poisoned arrows nor blow-pipes are used by the islanders, according to Mr. Man. These melancholy facts demonstrate that Mr. Holmes was not the paragon of Dr. Watson's fond imagination, but a very superficial fellow, who knew no more of the Mincopies (a mere nickname derived from their words for "come here") than did Mr. Herbert Spencer[1] ... but do not let us be severe on the great detective for knowing no more of anthropology than of other things! Rather let us wish him "good hunting," and prepare to accompany Dr. Watson and him, when next they load their revolvers, and go forth to the achieving of great adventures.

---

1 Herbert Spencer (1820-1903) is best known for having coined the phrase "survival of the fittest" in his *Principles of Biology* (1864).

# Select Bibliography

Baring-Gould, William. *The Chronological Holmes: A Complete Dating of the Adventures of Mr. Sherlock Holmes of Baker Street, as Recorded by his Friend John H. Watson, M.D., Late of the Army Medical Department.* Privately published, 1955.

——. *The Annotated Sherlock Holmes: The Four Novels and the Fifty Six Short Stories Complete.* London: John Murray, 1968, 2 vols.

Barsham, Diana. *Arthur Conan Doyle and the Meaning of Masculinity.* Aldershot: Ashgate, 2000.

Bengis, Nathan L. *The "Signs" of Our Times: An Irregular Bibliography of Arthur Conan Doyle's Novel "The Sign of Four."* New York: privately published, 1956.

Booth, Martin. *The Doctor, the Detective and Arthur Conan Doyle: A Biography of Arthur Conan Doyle.* London: Hodder and Stoughton, 1997.

Brantlinger, Patrick. *Dark Vanishings: Discourses on the Extinction of Primitive Races, 1800-1930.* Ithaca: Cornell UP, 2003.

Collins, Wilkie. *The Moonstone.* London: Tinsley Bros., 1868.

Ellis, Havelock. *The Criminal.* London: Walter Scott, 1890.

Frank, Lawrence. *Victorian Detective Fiction and the Nature of Evidence: The Scientific Investigations of Poe, Dickens and Doyle.* Basingstoke: Palgrave Macmillan, 2003.

Green, Richard Lancelyn and John Michael Gibson. *A Bibliography of A. Conan Doyle.* London: Hudson House, 1999.

Hall, John. *"I Remember the Date Very Well": A Chronology of the Sherlock Holmes Stories of Arthur Conan Doyle.* Romford: Ian Henry Publications, 1993.

Harris, Susan Cannon. "Pathological possibilities: Contagion and Empire in Doyle's Sherlock Holmes Stories." *Victorian Literature and Culture* 31 (2003): 447-66.

Klinger, Leslie S., ed. *The New Annotated Sherlock Holmes.* New York: W.W. Norton, 2005, 2 vols.

——. *The New Annotated Sherlock Holmes: The Novels.* New York: W.W. Norton, 2006.

Lang, Andrew. "On the Novels of Sir Arthur Conan Doyle." *Quarterly Review* 200.399 (July 1904): 158-79.

Lellenberg, Jon, L. *The Quest for Sir Arthur Conan Doyle: Thirteen Biographers in Search of a Life.* Carbondale: Southern Illinois UP, 1987.

Lycett, Andrew. *Conan Doyle: The Man who Created Sherlock Holmes.* London: Weidenfeld and Nicolson, 2007.

McBratney, John. "Racial and Criminal Types: Indian Ethnography and Sir Arthur Conan Doyle's *The Sign of Four*," *Victorian Literature and Culture* 33 (2005): 149-67.

Mehta, Jaya. "English Romance, Indian Violence." *The Centennial Review* 29 (1995): 611-57.

Mukherjee, Upamanyu Pablo. *Crime and Empire: The Colony in Nineteenth-Century Fictions of Crime.* Oxford: Oxford UP, 2003.

Pinney, Christopher. "Colonial Anthropology in the 'Laboratory of Mankind.'" In C.A. Bayly, ed., *The Raj: India and the British 1600-1947.* London: National Portrait Gallery, 1990, 252-63.

Redmond, Donald A. *Sherlock Holmes Among the Pirates: Copyright and Conan Doyle in America 1890-1930.* London: Greenwood P, 1990.

Sengoopta, Chandak. *Imprint of the Raj: How Fingerprinting Was Born in Colonial India.* London: Macmillan, 2003.

Siddiqi, Yumna. "The Cesspool of Empire: Sherlock Holmes and the Return of the Repressed," *Victorian Literature and Culture* 34 (2006): 233-47.

Spivak, Gayatri Chakravorty. "Can the Subaltern Speak?" In Cary Nelson and Lawrence Grossberg, eds., *Marxism and the Interpretation of Culture.* Urbana: U of Illinois P, 1988, 271-313.

Stashower, Daniel. *Teller of Tales: The Life of Arthur Conan Doyle.* London: Allen Lane, 1999.

Stashower, Daniel, Jon Lellenberg and Charles Foley, eds. *Arthur Conan Doyle: A Life in Letters.* London: HarperPress, 2007.

Tylor, Edward Burnett. *Primitive Culture.* London: Murray, 1871.

West III, James L. "The Chace Act and Anglo-American Literary Relations." In *Studies in Bibliography: Papers of the Bibliographical Society of the University of Virginia* 45 (1992): 303-11.

Wynne, Catherine. *The Colonial Conan Doyle: British Imperialism, Irish Nationalism, and the Gothic.* Westport, CT: Greenwood P, 2002.

## On the Indian "Mutiny" of 1857-58

Chakravarty, Gautam. *The Indian Mutiny and the British Imagination.* Cambridge: Cambridge UP, 2005.

Coopland, R.M. *A Lady's Escape from Gwalior and life in Agra Fort during the Mutinies of 1857.* London: Smith, Elder & Co., 1859.

Dalrymple, William. *The Last Mughal: The Fall of a Dynasty, Delhi, 1857.* London: Bloomsbury, 2006.

Grant, James P. *The Christian Soldier: Memorials of Major-General Sir Henry Havelock.* London: J.A. Berger, 1858.

Guha, Ranajit. "The Prose of Counter-Insurgency." In Ranajit Guha, ed., *Subaltern Studies II* (New Delhi: Oxford UP, 1983): 1-42.

Kaye, J.W. and G.B. Malleson. *The History of the Indian Mutiny of 1857-8*. London: W.H. Allen, 1888-89, 6 vols.

Misra, Amaresh. *Mangal Pandey: The True Story of an Indian Revolutionary*. New Delhi: Rupa & Co., 2005.

Muir, William. *Agra in the Mutiny and the Family Life of W. & E.H. Muir in the Fort, 1857: A Sketch for their Children*. np, 1896.

——. *Agra Correspondence during the Mutiny*. Edinburgh: T & T Clark, 1898.

Mukherjee, Rudrangshu. *Mangal Pandey: Brave Martyr or Accidental Hero?* New Delhi: Penguin India, 2005.

Thornhill, Mark. *The Personal Adventures and Experiences of a Magistrate during the Rise, Progress, and Suppression of the Indian Mutiny, etc.* London: John Murray, 1884.

Williams, Frederick S. *General Havelock and Christian Soldiership*. London: Judd & Glass, 1858.

## On the First and Second Anglo-Afghan Wars

The British Library, "A Chronological Survey of Afghan Sources," <http://www.bl.uk/collections/afghan/chronsurvsources.html>. Accessed 24 August 2008.

Havelock, Henry. *Narrative of the War in Affghanistan, 1838-9*. London: Henry Colburn, 1840, 2 vols.

Hensher, Philip. *The Mulberry Empire, or the Two Virtuous Journeys of the Amir Dost Mohammed Khan*. London: Flamingo, 2002.

Kaye, J.W. *History of the War in Afghanistan. From the Unpublished Letters and Journals of Political and Military Officers Employed in Afghanistan throughout the Entire Period of British Connexion with that Country*. London: Richard Bentley, 1851, 2 vols.

Malleson, George Bruce. *The Russo-Afghan Question and the Invasion of India*. London: Routledge, 1885.

Sale, Florentia. *A Journal of the Disasters in Affghanistan, 1841-2*. London: John Murray, 1843.

## On the Andaman Islands

Man, Edward Horace. *On the Aboriginal Inhabitants of the Andaman Islands*. London: Published for the Anthropological Institute of Great Britain and Ireland by Trübner & Co., 1884.

Mouat, Frederic J. *Adventures and Researches Among the Andaman Islanders*. London: Hurst and Blackett, 1863.

Mukerjee, Madhusree. *The Land of Naked People: Encounters with Stone Age Islanders*. New Delhi: Penguin, 2003.

Portman, Maurice Vidal. *A History of Our Relations with the*

*Andamanese. Compiled from Histories and Travels, and from the Records of the Government of India.* Calcutta: Office of the Superintendent of Government Printing, 1899, 2 vols.

Radcliffe-Brown, A.R. *The Andaman Islanders: A study in Social Anthropology.* Cambridge: Cambridge UP, 1922.

Sen, Satadru. *Disciplining Punishment: Colonialism and Convict Society in the Andaman Islands.* New Delhi: Oxford UP, 2000.

Tomory, David. *The Cannibal Isles.* London: Nthposition, 2003.

Weber, Georg. *The Andaman Association.* <http://andaman.org>. Accessed 24 August 2008.

Using 1339 lb. of Rolland Enviro100 Print instead
of virgin fibres paper reduces your ecological footprint of:

Trees: 11
Solid waste: 1,400 lb
Water: 11,078 gal
Air emissions: 3,640 lbs